The Principles of War
for the Information Age

Also by Robert R. Leonhard

Fighting by Minutes
Art of Maneuver

The Principles of War

for the Information Age

Robert R. Leonhard

PRESIDIO

BALLANTINE BOOKS • NEW YORK

For Bobby Leonhard, my son and best friend,
who consistently beats me in war games

A Presidio Press Book
Published by The Ballantine Publishing Group

Presidio Press is a trademark of Random House, Inc.

www.ballantinebooks.com

Library of Congress Cataloging-in-Publication Data

Leonhard, Robert R.
  The principles of war for the information age / Robert R.
Leonhard.
        p.   cm.
  Includes bibliographical references and index.
  ISBN 0-89141-647-1 (hardcover)
  ISBN 0-89141-713-3 (paperback)
  1. Military art and science—Forecasting. 2.War—Forecasting.
    3. Information technology. I. Title.
  U21.2.L397   1998
  355.02—dc21
                                        98-8287
                                        CIP

Printed in the United States of America

BVG 01

Contents

Foreword by Robert H. Scales Jr.   vi
Introduction by James J. Schneider   viii
Preface   xii
Part 1: The Framework for Change   1
1. The Mutation of War   3
2. Maryland, 1862: The Information Campaign   35
Part 2: The Principles of War   51
3. Maneuver   53
4. Offensive   80
5. Mass   94
6. Economy of Force   124
7. Objective   138
8. Security   162
9. Simplicity   170
10. Surprise   182
11. Unity of Command   194
Part 3: The Laws of War   205
12. The Law of Humanity   207
13. The Law of Economy   217
14. The Law of Duality   226
Part 4: The New Principles of War   241
15. The Arguments of War   243
16. The New Principles of War for the Information Age   251
Appendix: The Epistemology of the Principles of War   263
Works Cited   277
Index   283

Foreword

Over the centuries, military strategists and commanders have sought to decipher the complexities of warfare. They examined the great captains, hoping to distill their insights for future success. These insights became maxims that were further developed and refined during the Napoleonic era and then transitioned to industrial age warfare. By the mid-nineteenth century, they became known as the principles of war. For more than 100 years, these fundamental tenets have guided the U.S. Army. Despite massive changes in warfare over the last two centuries, these time honored principles have shown remarkable resilience.

The advent of the information age may change that. We are on the cusp of a revolution in precision. New and developing technologies will alter the future of warfare, providing tremendous increases in knowledge and speed. Knowledge enables us to know where we are, where our friends are, and where the enemy is. Speed will take the form of rapid deployment and high-tempo, pulsed operations that seeks to destroy an opponent's will to resist. The symbiotic relationship of knowledge and speed will allow the ability to maneuver under precision to attack targets that will cripple the opponent's center of gravity and destroy his will to resist.

But much about warfare will remain true to its past. War always has been a contest of wills. Centers of gravity will remain relevant. The key to future victories will be the collapse of the enemy's will to resist, while protecting our own troops. Time will remain a critical element in battle, and both sides will try to control it. Changes in the art of warfare will continue to follow technologically driven cycles. These traits of war are immutable.

Tensions in the continuity in warfare raise the critical question of whether the principles of war will remain valid under the conditions of information age warfare. Robert Leonhard has done all of us studying future warfare a great service by examining this issue in detail. His arguments are provocative; they confront long-accepted convictions. He challenges us to strip away preconceptions and to reexamine the principles of war in a new context. Dialogue and debate, especially in times of dynamic change, are indispensable for developing and refining ideas. From these debates spring the seeds of change.

Increasingly, our young army officers do not include themselves in the great doctrinal debates, nor are they challenged enough to investigate the principles which form the very basis of our profession. This is their opportunity to get in on the ground floor of a debate that could fundamentally change the way we fight. Lieutenant Colonel Leonhard is one of the few young officers who has stepped forward to join the debate, and he does so with enormous energy and intellect. Whether one agrees or disagrees with all of the premises in this work is secondary to the dialogue that it will generate—but that debate must take place. Only through a reasoned and vigorous give-and-take will we be able to refine the ideas that are vital to the continued evolution of our army as we prepare for war in the twenty-first century.

Robert H. Scales Jr.
Major General, U.S. Army
Commandant, U.S. Army War College

Introduction: The Cheat Codes of War

The eminent educational philosopher and psychologist Dr. Jean Piaget wrote that education is what you rely upon when you face the unknown. Lieutenant Colonel Robert R. Leonhard's new book is a major contribution to military education in Piaget's sense. The author offers a way to confront military problems of the future with innovative solutions.

This book is unique in at least two respects. First, as a contribution to professional education, the work goes beyond a "motor pool and rifle range" approach to topical military issues. Since the end of the Gulf War, the U.S. Army has moved from its post-Vietnam emphasis on education toward a renewed stress on training. Training gives education its practical significance; the two are opposite ends of the same continuum of learning. Both demand creativity, rigor, and insight. Training tends to be repetitive, rote, and methodical. Its purpose is to provide swift, responsive, and reflexive action in a deadly environment. Education, on the other hand, is reflective, integrative, and pattern-seeking. Just as training deals with the lethality of warfare, education confronts ambiguity. Bob Leonhard's contribution is to offer a new set of principles that can aid military leaders by imposing intellectual coherence upon an inherently complex and complicated phenomenon.

His second contribution is to analyze, critique, and recast the principles of war anew. The classical principles of war have too often been used as "cheat codes," a seemingly convenient list of ways to obviate the complexities and distressing paradoxes of armed violence. Leonhard is one of the few military theorists to make a comprehensive reexamination of the principles of war since their early expression in the 1920s by Major General J. F. C. Fuller; the other writer that comes to mind is former army colonel John Alger. What is unique about Leonhard's book is that he has reformulated the principles in light of the broader context of military history and the emerging potentials of the future.

THE HEURISTICS OF WAR
The principles of war are heuristic devices: rules of thumb that offer a quick entry into the solution of a problem. Instead of searching its entire knowledge base from scratch each time it tackles a problem, the mind employs ready to use rules of thumb that give swift and reliable access to solutions to a common set of problems. The principles of war provide just this kind of unique access. Nevertheless, as Leonhard reminds the reader, the principles do not provide a complete solution in the military decision making process, only the *beginning* of one. In the investment arena, for instance, a tried and true principle of stock trading is *buy low and sell high*. Behind this truism the investor must bring to bear a vast store of experience, knowledge, and understanding about the market before he invests even a penny. Principles, thus, are not substitutes for professional understanding, experience, and education. They can never become a mere checklist; for, as the late air force colonel John Boyd once said, "If you drop your checklist, your brains are below your feet."

Solutions to military problems are complex and convoluted, are seldom self-evident and may actually be contradictory. Even more troublesome for the military decision maker is the fact that warfare constantly evolves and reconstructs itself, mangling and ruffling itself in new and wondrous ways that are rarely evident.

If this is all true, then what? Why even bother with principles of war at all? Can we derive a set of principles that are heuristic and yet not destructive of creative solutions? Leonhard thinks so, and he offers a clear and compelling description of what those principles are—or should be.

THE DEAD HAND OF NAPOLEON
Why does Bob Leonhard want to reformulate the principles of war? In his view, they are no longer relevant to the study of modern warfare; indeed, they have not been entirely relevant since the American Civil War—over fifty years before their expression by J. F. C.

Fuller. As the author shows in his discussion and analysis, it is not enough to change here and there a principle or two. The principles of war evolved historically as a coherent whole developed to address a class of problems related to warfare; changing one principle without regard to the others wrecks the coherence of the whole framework. Changing one principle, therefore, *logically* entails changing and restructuring them all.

By and large, the principles evolved out of the experience of classical warfare—roughly the period from Alexander the Great to the defeat of Napoleon Bonaparte. The Industrial Revolution overturned the classical paradigm, but the old principles of war as problem-solving heuristics persisted in their out-dated formulation—their underlying logic remaining immovable, despite superficial changes in expression. To take one example, consider the Principle of Mass.

The Principle of Mass (or Concentration) ("Concentrate combat power and effects at the decisive place and time.") came from an interpretation of the Napoleonic Wars and received its fullest articulation in the writings of Baron Antoine-Henri Jomini. In the days of the concentrated, decisive battle, massing made eminent sense. But with the emergence of industrialized total war, this rule of thumb was overturned. Today, strictly applied, the Principle of Mass is logically contradictory and inconsistent with the other principles. First, Mass violates the Principle of Security ("Never permit the enemy to acquire an unexpected advantage."): given the lethality of today's weapons of mass destruction, concentration invites annihilation. Second, Mass subverts the Principle of Surprise ("Strike the enemy at a time or place, or in a manner, for which he is unprepared."): the foundation of surprise is deception, and a concentrated force is easy to find and hard to hide; find one enemy unit and you've found them all. Finally, Mass contradicts the Principle of Maneuver ("Place the enemy in a position of disadvantage through the flexible application of [massed] combat power."): a massed force is difficult—if not impossible—to sustain logistically for prolonged periods of time. Furthermore, massed forces are difficult to move swiftly and decisively to positions of advantage through a deep theater of operations.

These and other inconsistencies in the principles of war have compelled the author to take the sweeping and groundbreaking approach that you will find in the pages before you. In the last analysis, this book is an act of intellectual leadership. Bob Leonhard provides his reader with the intellectual purpose, direction, and motivation to question and challenge 175 years of military convention and usage. Who among you will follow?

James J. Schneider
Professor of Military Theory
The School of Advanced Military Studies
Ft. Leavenworth, Kansas

Preface

Practice, without theory and reflection,
dwindles into unsatisfactory routine.
—Frederick the Great

This is not a book to be agreed with. It is a book to be argued about.

It is difficult for me to state this, because I am firmly convinced that the conclusions in this book are accurate. However, that belief does not automatically translate into a desire on my part that all readers believe as I do. This is, after all, only one soldier's judgment on the character and direction of future conflict in general, and armed conflict in particular. But what I do hope to achieve is to stimulate rigorous debate: a commodity that is conspicuously absent today.

It's not that we are stagnating in our efforts to build a viable national strategy or its military component. Indeed, change proceeds at an alarming rate. The services are continuing to assimilate information technology, and the reorganization of our armies, fleets, and air wings likewise progresses at a breathtaking speed. In a furious pursuit of what was once called "the peace dividend" following the Soviet collapse, Congress and the National Command Authority have cut deeply into the armed services, causing a serious reassessment of how we will continue to defend our interests.

This book is *not* an appeal for bigger defense budgets, nor a reactionary attempt to return to the glorious Cold War days. I am not motivated by defense politics, nor by the fleeting issues of the day. This is a book about the nature of conflict in the twenty-first century.

I use the term "conflict" intentionally, in order to widen the scope of the book beyond a merely military context. I am not a businessman, yet I perceive that many of the points we will discuss in this book are applicable to economic conflict. In fact, in a broad sense, the principles that I will draw out in the pages ahead should bear fruit in all aspects of human conflict: business, politics, sports, social conflict . . . in some sense, even in love and romance—a human sport built around the most sensual conflict of all. But most of all, the principles within this book are aimed at military conflict.

xii

It has been my goal from the start of this project to write a work that will last. One cannot read the classic works of Sun-tzu, Vegetius, Saxe, Clausewitz, or Brodie without at once seeing the restricted context in which their conclusions and wisdom pertain. Some past writers—Xenophon is a good example—seem completely absorbed with the transitory problems of their day, and almost totally sidetracked by the technological state of the art. To the degree that other writers were able to lift their gaze beyond their own place and time in history, they were more successful and long-lasting in their impact.

I believe that it is easier today to separate oneself from the immediate technological and political context, because we are a generation used to rapid change. Within the small space of my own military career, I have watched as my profession leapt from World War II-era mechanization into a world of digital communications, computers, and precision weaponry.

With this advantage, I have set myself to compose a work on the principles of war that will pertain not only to our own but future generations as well. While all writers are subject to their own set of assumptions and prejudices, my intent is to minimize them so as to address those factors of conflict that apply to *all* the ages, both past and future.

It is my hope that readers will find this book both entertaining and useful, and that they will apply these ideas to whatever conflicts come their way. Nevertheless, I have chosen to write from the perspective that I know best: I am an American army officer. I have written mostly about warfare, trusting, however, that readers will apply these principles to other aspects of life. I have written mostly about the United States Army, although I believe my colleagues in the other uniformed services will find these ideas useful not only in land warfare, but also in the air and on the seas. I have written as an American, but with the sincere desire that my friends in other nations can take these principles and apply them with success to their own particular challenges. In other words, I have written from a narrow, personal perspective— for the purpose of producing a work with broad applicability.

I began this preface by claiming that the doctrinal arguments which should underpin change lack rigor. We do not lack for articles, white papers, and concepts concerning the future of the armed forces, but *quantity* does not suffice. It is the *quality* of the argument that is the problem. In my experience within the service, as well as in my reading, I perceive an unflagging devotion to traditional ideas, most of which are hopelessly and dangerously outdated. And—what is worse—a seeming reluctance even to question those ideas.

Under the inspired leadership of the Army Chief of Staff and the Training and Doctrine Command, the United States Army, for example, has been conducting the most forward-looking experimentation in history. Under the rubric of "Force XXI," the Army has committed itself to investigating the phenomenon of Information Age warfare. Through both live and computer simulation, we have been summoning the most futuristic and effective land force conceivable—integrating the latest advanced weaponry with digital command and control. And if that were not enough, agencies in the Army are concurrently looking at the next step beyond Force XXI. We are simultaneously building the future Army and envisioning the Army beyond that. In all these considerable efforts, as well as within the joint services, well-intentioned officials are thinking seriously about tomorrow.

The consequent debates are lively. Experts challenge the feasibility and suitability of developing technology. Others argue about whether future warfare will feature recognizable clashes between organized forces, or patternless acts of violence erupting without the benefit of political control. In many dimensions, the debate about the future continues.

The problem is that it is not *rigorous.* It has yet to seriously challenge basic beliefs and gut-level issues. Military and civilian leaders are still in their comfort zone concerning the character of future war. We need to break into the temple and smash a few idols if we are to summon the future with true conviction.

No armed force in the world is well designed for debate. The hierarchical organization of armed forces is designed at best to *control* debate, and at worst to *prevent* it. The strength of the military hierarchy is easily underestimated by the outsider. But within the services, it remains simply unthinkable to disagree (in a big way, at least) with the flag officers. New ideas are stillborn unless a general attaches his or her name to it. And flag officers, for better or worse, do not argue with each other . . . at least not in peacetime.

When the senior officers avoid serious disagreement, they ipso facto squash meaningful debate elsewhere in the services. The result is a contained, controlled, and carefully directed monologue that is short on drama, and long on consensus.

This is not all bad. Disagreement does not, by itself, accomplish anything. Sooner or later, budget decisions must be made, manuals written, and materiel developed. It takes a huge team of people—both military and civilian—working together to accomplish change, and the military hierarchy does well at unifying action.

But somewhere in the process of change, we owe it to ourselves to disagree—with passion, conviction, and anger, perhaps—but most especially with a disciplined intellect. If we fail at the beginning to challenge each of our basic assumptions, we are only delaying and magnifying the problem for someone else to solve.

That, then, is the purpose of this book: to generate serious and rigorous argument. To put forth a strong thesis that begs an antithesis . . . and eventually a synthesis.

The best part about writing, like traveling, is the people you meet. I have had the great fortune to work with many folks on this project and to benefit from (in some cases, steal) their excellent insights.

Dr. Lee Forester, a valued friend, had both the intellect to wade into this effort, and the guts to confront me when I needed it. Were it not for his time and expertise, I could not have finished this book. Together, we wrestled Chapter Four to the ground, and his wisdom and insights are found throughout this text. "Faithful are the wounds of a friend . . ."

Dean Essig and all my friends from the Gamers put the ideas in this book to the test using the best war games on the market and gave me merciless (and valuable) feedback. Thank you for your profound expertise in military history, and for your attention to this project.

Lieutenant Colonel Rob Barry, with whom I have engaged in hundreds of hours of debate since 1994, turned his considerable talents loose on this effort and helped me shape many of the ideas in this book.

Colonel Dan Bourgoine, longtime friend and counselor, used his years of experience and operational expertise to redirect my steps when I went astray.

Lieutenant Colonel John Stawasz and Capt. Mike Pryor provided much needed guidance and advice.

Major Eric Walters, USMC, instructed me concerning crucial aspects of intelligence and command doctrine that I had missed. Sincere thanks for our many hours of e-mail discussion.

Major General Robert H. Scales Jr. soldier and intellectual, gave me the benefit of his immense experience and wisdom concerning the future of the army, as well as the foreword for this book.

Dr. James J. Schneider, professor and mentor to me and to a generation of officers at the School of Advanced Military Studies, gave me encouragement and his infinite wisdom and knowledge on the subject, along with an introduction.

Robert Kane, Richard Kane, and E. J. McCarthy, along with the rest of the team at Presidio Press have made writing this book a true joy and a treasured professional experience.

Colonel R. B. Thieme, Jr., my pastor, and all my friends at Berachah Church in Houston, showed me hospitality and grace, and allowed me to engage in a never-ending discussion concerning strategy and policy. Many thanks!

To my children, who missed some Daddy Time while I worked on this project, my love and gratitude.

Most especially, my undying thanks and devotion to my partner and wife, Carol—the other author of this book—who sustained this project with her energy and intellect.

PART 1
The Framework for Change

We still persist in studying a type of warfare that no longer exists and that we shall never fight again.

—Roger Trinquier, <u>Modern Warfare</u>, 1961

1: The Mutation of War

Don't worry, boys, you'll be able to go over the top with a
walking stick. Not even a rat will have survived.
> —General Rawlinson to British troops before
> the Battle of the Somme

We live in an age of proliferating information and shrinking sense.
> —Jean Baudrillard

This is the very limit of endurance . . .
> —British soldier during the Battle of the Somme

On 1 July 1916, Gen. Douglas Haig ordered two British armies out
of their trenches and into the teeth of a prepared German defense
near the Somme River. By the end of the day, 19,000 British soldiers
were dead, and another 40,000 were wounded or missing. Haig, an
experienced veteran and a highly educated officer, had planned and
directed the worst single-day loss in British history.

By the standards of the day, the First Battle of the Somme was reck-
oned a marginal success. It forced the Germans to shift reinforcements from the threatened Verdun front. Yet by the end of the
Somme operation, 1,200,000 soldiers from all sides (British, French,
German) had become losses, while the front lines barely shifted at
all. And despite the horrifying loss of life, the war dragged on for
two more years. From the perspective of the profession of arms, Haig
and his fellow officers on both sides of the trench lines had utterly
failed.

The human lives lost during those years have passed from personal memory into the history books. The ravaged bodies of the soldiers who followed orders into an inconceivably brutal machine of

death have decayed. And the passing of years has seen even more violence and tragedy, each episode diverting our attention and breeding new generations of widows and orphans. Yet the bloody spectacle of World War I must be of enduring interest for the student of warfare, because the cause of so many deaths in that war was singular. Ambition, greed, animosity, and fear all played their part, of course, but the single most brutal slayer in the First World War was *ignorance.*

Men died out of proportion at Ypres, Flanders, Passchendaele, and Verdun. We must leave it to philosophers to judge whether a given strategic objective is worth the lives of one, ten, or ten thousand soldiers. But there is no doubt that the destruction of human life from 1914 through 1918 exceeded all bounds of proportionality, regardless of one's philosophic perspective. There was too much death in World War I.

Why did this happen? Why did Douglas Haig and his contemporaries oversee such a profligate harvest of souls? The fundamental problem was *change.* The character of warfare had changed during Haig's lifetime, and in fact, in the space of his active career. Technological advances had intruded into the military art and utterly transformed it. Machine guns, rifles, rapid-fire artillery, and other advances had arrived and demanded a serious reassessment of the military art.

The lessons of the recently concluded Boer War and the Russo-Japanese War were studiously misapplied by Haig's generation. Despite the portents of smokeless gunpowder, breech-loading rifles, improved artillery, and machine guns, leaders in World War I chose to magnify anecdotal episodes of classical maneuver, rather than see the larger truth that infantry charges were becoming horribly inappropriate in modern war. It was time for the leaders to reconsider their profession. Instead, most soldiers of the day decided that, in the words found in the introduction to Foch's *Principles of War:*

> The present war has, in spite of all its novel features, shown once more that the fundamental principles of tactics remain unchanged . . .

Were they correct? In view of the dreadful loss of soldiery during those years, we must conclude one of two possibilities: Either the principles of war *had* changed and the leaders simply failed to notice, or the principles remained immutable but adherence to them did not lead to advantage in combat. If the latter is true, then the principles of war are merely academic and not worthy of study by the military professional. Today's military leaders are too busy to devote themselves to truisms that have no practical value.

The weight of historical evidence, as well as the force of common logic suggest otherwise: There *are* practical principles that can guide leaders to success. But those principles change from age to age. They are not immutable, because the factors that influence the conduct of armed violence are constantly evolving. The great failure of the generation of military leaders in World War I was their refusal (with notable exceptions) to adapt quickly to change.

Douglas Haig himself considered the horrendous casualties of the war quite justified, when he wrote his last official dispatch in March 1919. Commenting on the Somme Offensive, he wrote:

> In every stage of the wearing-out struggle losses will necessarily be heavy on both sides, for in it the price of victory is paid. If the opposing forces are approximately equal in numbers, in courage, in moral[e] and in equipment, there is no way of avoiding payment of the price or of eliminating this phase of the struggle.

It is puzzling that Haig could conclude the necessity of an extended attrition phase in modern war, when he had himself witnessed the startling collapse of his own lines in the face of innovative German tactics in the Spring Offensive of 1918. Using infiltration tactics combined with new ways of deploying and controlling artillery, the Germans tore huge gaps in the Allied lines before their own lack of planning and resources brought their successes to an end. But according to Haig's biographers, he had formulated this idea of a "wearing-out phase" long before World War I. Even when faced with palpable evidence that there was a better way to fight,

Haig's presuppositions convinced him that bloody attrition fighting was an inescapable part of war.

Doctrinal thinking eventually caught up with the technological state of the art, but not until the Battles of the Somme, the Aisne, Ypres, and others destroyed a generation of European men. Was this a justifiable cost? Can knowledge of warfare progress only through the effusion of blood and the loss of so many lives? Is it not the duty and moral obligation of military leaders to think about war and adjust military doctrines in anticipation of such dramatic changes?

If the military profession cannot answer these questions, then it ceases to be an art, a science, and a legitimate profession. There is no moral right to command apart from the responsibility to protect the lives of the led. War will always result in the deaths of some, but the profession of arms demands that the leaders do everything possible to avoid needless death. The urgency of these issues is upon us today, because the nature of warfare has changed once again. And this time, the mutation of war is far greater than anything yet witnessed by man.

Unfortunately, it is a favorite dictum within the U.S. armed forces (and the army in particular) that "doctrine drives technology." In other words, it is commonly preached by modern soldiers that doctrinal thinking and writing should determine the development of military technology. This idea is an expression of the soldier's natural distrust of the scientist, and, even more so, of the ineffable conservatism of military professionals. Unfortunately, the historical record is clear in its repudiation of "doctrine driving technology." In fact, just the reverse is true.

What came first—cavalry charges or the stirrup? Blitzkrieg or the internal combustion engine? Airpower doctrine or the airplane? In these and countless other examples, it was the technology that came first . . . followed ever so slowly by doctrinal thinking. Soldiers come kicking and screaming into revolutionary changes—often led there by more perceptive civilians—all the while insisting loudly that "Doctrine drives technology!" despite clear evidence to the contrary.

The pattern is about to repeat itself, because military dogma notwithstanding, technology has once again changed the military art.

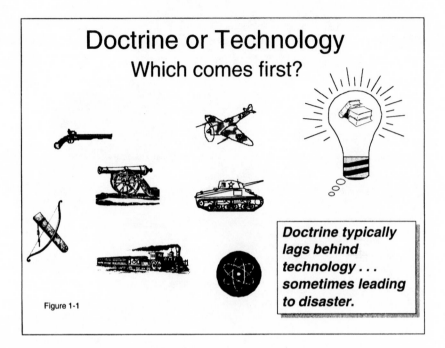

Figure 1-1

Warfare is an outgrowth of the human soul, and the ironic, para-doxical, sometimes chaotic nature of the soul inflicts upon the mil-itary art a maelstrom of dramatic contradictions, fleeting insights, and ever-changing truths. Our business is redefined every day. Some-times we notice; more often we do not.

Successful navigation through the buffeting winds of the military art requires a peculiar combination of stable doctrines and the most audacious flexibility. In this context, there is a list of nine or so words that have been the center of controversy and argument—and some-times the cause of the most maddening consensus—throughout the history of the world and of this nation. Together, these words are known as the principles of war.

It is disturbing that, in a display of the grossest pedantry, we per-sist within the American armed forces in our insistence upon the im-mutability and applicability of these principles, which were devel-oped many years ago. Even when these truisms are demonstrably

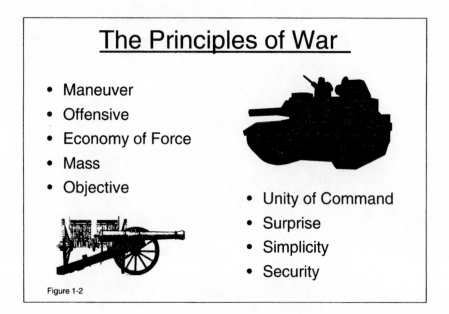

The Principles of War

- Maneuver
- Offensive
- Economy of Force
- Mass
- Objective

- Unity of Command
- Surprise
- Simplicity
- Security

Figure 1-2

inaccurate, military officers cling to them beyond all reason—choosing more often to reinterpret reality than to modify these ideas.

We point to successful armies that didn't mass, and we claim that they "massed effects." We consider victorious warriors who won while on the strategic defense, and we point to their occasional use of offensive tactics to prove the efficacy of "Offensive." We view the enormous complexity of Operation Just Cause or Desert Storm and yet claim that they were true to the principle of "Simplicity." We point to disunited, successful armies as proof of "Unity of Command." We permit the most dull-witted frontal attacks to prove "Maneuver," . . . as long as they work.

There are compelling reasons to revise the principles of war. The historical record is clear: The principles are neither unchanging nor universally accepted. They have in fact changed many times even in the brief history of our country. Other nations—some our close allies—disagree with our list of principles, some substituting their own lists, others claiming that there are no valid principles. We owe it to ourselves, therefore, occasionally to reexamine our beliefs. Principles are developed in schoolhouses, not handed down from Mount Sinai. Regardless of our prejudices or our conclusions, we can benefit by a rigorous scrutiny of the principles. Either we will graduate to a newer, better way of thinking about war, or we will confirm for ourselves that the current list is about right. Either way, we have thought critically about what is literally a matter of life and death.

What would happen if a modern politician stood and delivered a speech on the utter necessity of strict laissez-faire economics in modern America? Or if a prominent theologian espoused a reversion to a divinely ordained monarchy as an effective form of government? What if a prestigious scientist put forward the proposition that rats spontaneously generate from garbage? Surely, we would gape in disbelief or roar with laughter at these outdated ideas. Why then do we bestow dignity and unquestioning awe upon military ideas from the distant past?

It borders on the astounding that so many military professionals as well as civilians accept the principles of war with little reservation or skepticism. A cursory look into the development of some of the most time-honored ideas that comprise the principles will find historical contexts that are completely foreign to us today. Ideas about mass, objective, offensive, and others come from periods in history in which monarchs conducted wars of conquest; democratic liberalism was nonexistent; and technology had not progressed significantly beyond that of ancient Rome. If we were to try to import social, political, or economic ideas from those days into modern society, we would become objects of ridicule. Yet, almost unthinkingly, we welcome the aged, enfeebled principles of war onto the battlefields of tomorrow. This blind acceptance is, at best, self-delusion.

Further, the principles of war as we have them today were little more than battlefield techniques when first conceived many years ago. When first expressed by classical ancient writers, ideas about mass, maneuver, and security centered around the *tactical* level of war—that is, they addressed what commanders should do when in direct contact with enemy forces on the battlefield. Today, we consider these same ideas to be applicable at the operational and strategic levels of war as well. It is as if we believe that a *war* is nothing more than a *battle* that has gotten out of hand. It is like a father teaching his son that principles of boxing will also serve as reliable principles of life. Before the late nineteenth century, tactical and strategic ideas were most often indistinguishable from each other. If we were to resurrect some of the great thinkers from whom we derived the principles of war, and ask them whether those principles related more to tactics or strategy, most would not understand the question. Until very recently, military strategy was, for the most part, military tactics writ large. The classical principles of war began life as principles of *battle*.

But beyond a general consideration of the principles in the light of history, we have an obligation to revisit them as we emerge—both chronologically and technologically—into the twenty-first century. Things are happening around us in the realm of politics, culture, and science that, regardless of entrenched military conservatism, will overturn aged and respected beliefs. The world is changing, and the soldier who is loyal to his nation and serious about his profession must overcome the anachronistic thinking that follows modern armies like the merchants and prostitutes who dogged the footsteps of armies of the past.

The purpose of this book is to examine each of the principles of war and to comment on their validity and utility. We shall consider each of the nine currently embraced by the U. S. Army not only for their applicability in history, but also in light of recent changes in warfare. Many have claimed that we are in the midst of a "revolution in military affairs." (I, for one, believe we are.) If so, then surely we owe it to ourselves and to posterity to challenge the most fundamental expression of our military thought.

In a general sense, we can categorize the existing principles of war into two groups according to the underlying logic. First, there are principles of *convergence*. That is, of the nine principles, five of them have a common logic: *reducing diversity to unity*. Mass, Objective, Unity of Command, Economy of Force, and Simplicity are all principles of convergence. Mass instructs us to converge *combat power* toward a single point in space and time. Objective moves us toward one *purpose*. Unity of Command advises us to have one *commander*. Simplicity (the word itself is a cognate of "single") pushes us to reduce com-

CATEGORIZING THE PRINCIPLES OF WAR

Principles of *Convergence*

Mass
(One <u>point</u> in space and time)
Objective
(One <u>purpose</u>)
Unity of Command
(One <u>commander</u>)
Simplicity
(One <u>idea</u>)
Economy of Force
(Minimize divergence)

Principles of *Interaction*

Offensive
(One side <u>attacks</u> the other)
Maneuver
(One side <u>dislocates</u> the other)
Surprise
(One side <u>preempts</u> the other)
Security
(One side <u>forestalls</u> the other)

Figure 1-3

plexity to one *idea*. And Economy of Force—viewed traditionally as the converse of Mass—urges us to avoid wasting resources on divergent activities. Each of these principles emanates from the logic of oneness.

The other four principles have to do with *interaction* between the enemy and friendly forces. Each of these—Maneuver, Offensive, Surprise, and Security—envision an unequal relationship between the two opponents: That is, one side is supposed to do something to the other side. In this regard, these principles of interaction instruct us that all interaction between enemy and friendly forces should be unequal: One side should do bad things to the other. Maneuver tells us that one side should *dislocate* the other. Offensive informs us that one side should *attack* the other. Surprise calls for one side to *preempt* the other. And Security urges one side to *forestall* the other.

These four ideas are founded on a common logic: Interaction in war should be unequal, favoring the friendly side.

As a general rule, the principles of convergence are invalid in the Information Age. The logic of oneness does not work in the twenty-first century—and in many cases did not work in the past, as we shall see. The logic behind the principles of interaction still pertains, but most of these principles are poorly framed and lead to grave misunderstanding concerning the true nature of interaction on the battlefield. As a result, even these principles of interaction must be revised if they are to retain their utility.

THE CONTEXT OF TWENTY-FIRST CENTURY WARFARE

We begin our journey to rediscover warfare by proposing that there is a need for reassessment. But why? What has changed?

We must tread lightly as we answer that question. On the one hand, we must avoid pointing to transient developments. No doubt the first soldiers who employed dirigibles stood amazed as the gigantic airships ascended over the battlefield. Yet history did not christen the early twentieth century "the Blimp Age." Instead, other technological developments quickly relegated dirigibles into the Curiosity Shoppe of the ages. It is in our nature to be easily awe-struck by new gadgets, and, in the course of my military career, I have done my share of gaping. But the history and theory of armed combat are not so readily amazed, and we must perforce be circumspect, looking for *trends* instead of gadgets.

To complicate our vision of future warfare further, the strategic context of the next century is difficult to predict. Prognosticators abound, however, and it becomes our unenviable duty to sort through them all for some glimmer of truth. Some predict a future world in which nation-states whither away before international cartels and hordes of angry, ungovernable fanatics. Others contribute apocalyptic visions of disease, starvation, and global warming—all of which combine to overtax our anachronistic concepts of government. Still others point to a resumption of the Cold War, this time spiced with insidious nuclear terrorists. A variation of this theme proclaims optimistically that the twenty-first century will be the American Age, during which the United States of America continues to call the shots. There are even a few who foresee (as so many of their core-ligionists have throughout history) the end of warfare and the beginning of the Age of Peace.

Against this stupefying array of strategic and technological imponderables stands the equally dangerous bulwark of conservatism. The ineffable power of doing nothing is a force often overlooked in the history of mankind. Inertia is a cold-blooded killer. Yet, when attended to carefully, it sometimes offers nuggets of wisdom.

It is along this difficult high wire that we choose to tread in this book. We shall consider both the factors of change, and those of stability. We shall seek solutions that will prevail in conflict scenarios, regardless of which future comes our way. And, in the end, we shall retire some highly respected but utterly false concepts of war, while advancing a set of new ideas about how to prevail in warfare.

But first we must outline what exactly has changed about human conflict in general, and warfare in particular, for our basic assumption is that a significant mutation of the military art is indeed ongoing.

The Information Age
For the past sixty years, and especially since the end of the Cold War, observant people from diverse backgrounds have noticed a qualitative change in the culture of (at least) Western man. This change has eventuated in the collective effort to name a new age in history. The period—defined most directly in terms of technology—has variously been called the Information Age, the Technology Age, the Postindustrial Age, the Computer Age, and the Third Wave.

More recently, in the wake of the dramatic collapse of Soviet communism, Western thinkers have begun to apply the idea of a qualitatively different era to the study and practice of warfare. As a result, military periodicals and books have been replete with articles and essays on "Information Age Warfare." The problem is that the study of the military aspects of this advance in technology has been conspicuously lacking any sociological, philosophical, or theoretical component. Perusing the various writings on information war available today, one would think that by understanding the technological components, one has embraced the totality of this new phe-

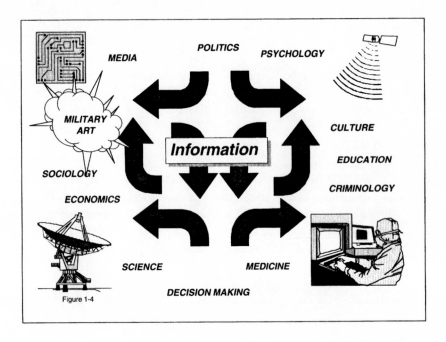

Figure 1-4

nomenon. Such a conclusion would, of course, be absurd. We can no more understand war by contemplating hardware than we can understand a parade by observing a foot. The technology may be a vital component of information war, but divorced from its human context, it does not begin to explain the problems, challenges, or potentials of this new concept of conflict.

In World War I the challenges and dilemmas that occurred in command and control were incomprehensible to those who had examined only the communications hardware developments beforehand. When neatly separated from their sociological context, the telegraph, telephone, and wireless communications seemed to solve myriad problems for contemporary soldiers. But when overlaid on the chaotic politics of Europe and the foot-weary, blood-drenched soldiery locked in combat in Picardy, these new instruments of communication caused as many problems as they solved. In effect, they separated the armies from their generals, both morally and physically, with the result that operations degenerated into indecisive stagnation—a stagnation completely incomprehensible to the absent leaders. In a similar manner, soldiers have failed throughout history to discipline the assimilation of new technology with intellect and reflection.

The terms describing the present age are by no means synonymous, nor do they necessarily point to the same span of years. As we will see, the "Computer Age" fits within the "Information Age," but the latter is a much more inclusive term. The "Third Wave," a term coined by Alvin and Heidi Toffler, is primarily an economic classification of time—but one with significant sociological implications. Likewise, the concept of "Postindustrial Society" is economic rather than technological, but it also impacts logically on culture and military science. We could perhaps claim to be living in the "Technology Age," based on the rate of technological change, or the degree to which technology penetrates into everyday life across socioeconomic boundaries.

Clearly the philosophical efforts to classify the current age of mankind will come to fruition only when all of these new trends pass into the history books. After all, the Dark Ages could be viewed as dark only from the light of the Renaissance and the Enlightenment. But for the purposes of this book, we will employ the concept of the Information Age as being fundamental to the changes taking place within military art and science.

What and when is the Information Age? Although we can no doubt come close to a conceptual understanding of the idea, it would be an impossible undertaking to specify the moment in human history when man progressed into the new era. In a sense, the age of information began more than five thousand years ago with the invention of writing. Writing permitted the communication and storage of information through other than oral means. The invention of writing also had a profound effect on society. Whole professions grew up around the ability to read and write, just as today, computerization requires specialized professions. Further, the storage of information on rock, papyrus, or paper permitted an edification effect on human knowledge. In other words, each generation could rapidly assimilate old discoveries and use them to discover yet more. In a real sense, writing allowed each generation to stand on the shoulders of the one before.

The technology of writing and storing information progressed throughout human history, but the most important step forward came long after the first cuneiform was carved: the printing press. When Johannes Gutenberg perfected movable type in the 1440s, he endowed mankind with the ability to transmit information to innumerable people, without the danger of that information being corrupted in transmission. In short order, the printing press boosted the development of science and learning in general.

The invention of the printing press radically changed the course of human history. The Protestant Reformation, in one sense, was a revolution of the printed word against the written word. Journals and newspapers proliferated, and books became available by the thousands. The social impact of print—an impact with far greater yield than the most powerful nuclear weapon—was profound. Still, in order to convert the potentials of information into hard reality, man would have to await the Industrial Revolution and its attendant phenomena of mass literacy and a popular press.

Even given the sociological impact of the printing press, history has not chosen to call the fifteenth century the Information Age. This name was reserved for the present day. The origin of that name can perhaps be found in two specific developments: television (and mass media in general) and the computer.

The development of the radio in the early twentieth century and of television in the 1940s created another sociological revolution. The sudden ability to broadcast a message to millions of receivers soon led to new methods in politics, business, and of course, military operations. The high-tempo mechanized operations that characterized the most dramatic moments of World War II relied on wireless communication to facilitate command and control. When television came into use in the mid-twentieth century, it impacted squarely upon the relationship between the government and the governed. In the decades since, the television camera has scrutinized America's wars and, by exposing warfare's true nature to the world, has changed the way we fight.

The computer is also central to Information Age technology. Some have suggested that the computer was invented just in the nick of time . . . just as Western man was about to drown in the proliferation of information. Mass media, the growth of the white-collar sector of the economy (which overtook the blue-collar sector in terms of overall numbers in 1956 in the United States), and the post–World War II economic boom led to a critical need to process ever more complex and numerous data. Fortunately, the computer stepped into the breach.

Computers have become much more than data processors in the past two decades. They are, in fact, an integral part of life in modern countries. From digital wrist watches to complex targeting arrays, the computer has invaded every aspect of Western man's life. It runs machinery, aids in decision making, talks to other computers, and stores data. But the computer is a mixed blessing. Although it certainly filled the growing void between available data and human processing speed, the computer also added to the information burden. It is not simply a *processor* of information; it is also a *producer.* Man invented the computer to carry his information-processing load, but shortly after he sighed with relief, he discovered that the computer was a sort of breeder reactor of information: it consumed, but it could also produce more than it consumed.

This dual nature of the computer leads directly to a fundamental, perhaps philosophical, dichotomy of thought concerning the Information Age. One school of thought maintains that the computer has narrowed the gap between the mind of man and a *finite* set of data (i.e., "all that is knowable"). Hence, the computer is a revolutionary tool that provides man a way of coming to total comprehension of the world around him. The other school of thought suggests that the information *produced* by the computer is greater than that *processed* by it, so that the gap between man and total knowledge is in fact growing. The logic here is that by processing two data points and then combining them through some synthetic procedure, the computer allows for infinite permutations of data. The question, in short, is whether man is becoming more or less ignorant. We find this issue on the battlefield today, as commanders and staffs struggle with information overload.

Other technological tools also relate to the Information Age. Sensor technology is a growing industry. One of the salient features of the U.S. armed forces is the ability to detect hostile military operations through exploitation of the electromagnetic spectrum. The most widely publicized advances in this area include the Joint Surveillance and Target Acquisition Radar System (J-STARS), an air force/army radar system that operates from an E-8 aircraft. Employing a moving-target-indicator radar array, J-STARS enables a theater or lower-level commander to surveil an entire theater of operations and target moving enemy formations.

In addition to both ground and air radar, military commanders across the globe can also make use of satellite imagery. As satellite networks grow, this expansive source of information can provide detailed, accurate terrain and weather maps as well as real-time imagery of enemy forces and installations. By exploiting various portions of the electromagnetic spectrum, satellite-borne sensor arrays can investigate terrestrial phenomena that would be undetectable along the visible spectrum.

Soldiers and weapon systems likewise benefit from advances in sensor technology. Early image-intensifying night optics have progressed through second-generation thermal technology and beyond, permitting an ever-greater capability to detect, classify, and identify targets on the battlefield through darkness and obscurants. By integrating laser technology as well, soldiers can also determine more precisely where a target is.

As sensor technology advances and integrates fully with command and control processes, it moves us into a new way of planning and conducting operations. In the past, armies have, for lack of a better term, conducted "estimate-based" operations. The U.S. Army, as an example, plans operations based on sophisticated guesses at where the enemy is and what he might do. Our intelligence doctrine and processes—arguably the best in the world—are inextricably bound up with this idea: We are fundamentally ignorant of the enemy's whereabouts and intentions, and so we *estimate* the future. In the past, there simply was no other way to do business, and the hallmark of a great commander or staff was how well (i.e., quickly and accurately) they could estimate the enemy.

Likewise, estimates characterized every part of staff preparation. Not only do staff officers estimate what the enemy is doing, but also they must assemble similar predictions on *friendly* strengths and dispositions. Nor are estimates confined to prognostication: Commanders and staffs must also estimate friendly and enemy *current* statuses, because even the present disposition of one's own troops is often unknown. Underlying every good operation order is a logistical estimate, a personnel estimate, an operations estimate, and so on. Current military doctrine is "estimate-based."

But what happens when technology allows the commander and staff to transcend estimates and replace them with truth? What might follow if future leaders can finally harness that most elusive prey, *knowledge*, and be able to reliably understand the battlefield even as events are unfolding? On the one hand, we can avoid this difficult question on grounds of technological feasibility. It takes no effort, after all, to propose that nothing changes. But if we are willing to look this potentially revolutionary development in the face, we will see that when truth-based planning replaces estimate-based planning, at a stroke the art and science of war change dramatically.

Of course, arrayed against any technological advantage are attempts to counter the advantage or defend against it. Just as an opposing nation would seek to improve its defenses against a new missile, likewise we can count on the enemy to try to offset our advances in sensor technology. Indeed, it is characteristic of military technological developments that an army will try to develop counters to its own technological advances. Concurrent with advances in the ability to detect, locate, classify, identify, track, and assess damage against enemy targets, other technologies aim at denying these activities. "Low-observable" or "stealth" technology represents an evolution of capability that began with the first primitive attempts at camouflage. Modern technology, however, has attained sophisticated (albeit expensive) methods of hiding from enemy observation. Essentially, stealth technology aims at reducing the distinctions between an object and its background, just as sensors attempt to emphasize and increase that distinction.

Information Age warfare will see these trends in sensor and countersensor technology continue. In ancient warfare, the detection of enemy targets was less of an issue than the destruction of them. As warfare progressed through the Industrial Age into the Information Age, however, armies dispersed on the battlefield and sought to hide their activities to prevent enemy fires. The relationship between detection and destruction has in our day reversed: Accurate detection of the enemy and security against enemy detection of friendly activity is the theme of modern warfare. The actual destruction of detected targets becomes a *fait accompli*—an inevitable sequel to gaining accurate information.

Related to the idea of stealth technology, armies that are attempting to hide from enemy fires may also choose to fade into the *sociopolitical* background. Without doubt, the future U.S. military will face the tough challenge of fighting enemy armed forces within crowded urban areas. By placing themselves in close proximity to noncombatants, enemies attempt to dislocate our fires. In a sense, this is an attempt to use political leverage against our military advantages. The military response, of course, will be aimed at isolating legitimate targets from the noncombatants, and information technology has a big part to play in that effort.

Attending all these efforts must be a general revitalization of thought concerning urban warfare. The American biases and preconceived notions about city fighting are the primary roadblock toward effective developments. Because our thinking is built more around the Battle of Stalingrad of World War II instead of the more relevant and recent battles in Mogadishu and Grozny, we have a defense community that is ill equipped to prepare for future urban warfare.

It is a common misconception that urban fighting requires large amounts of light infantry. Nothing could be further from the truth. The force of choice for most future urban scenarios will be armored, mechanized forces. The notion that we need large dismounted forces comes from a false estimate that future urban fighting will feature the systematic clearing of rooms and buildings as the centerpiece of tactical operations. Such fighting is, in reality, but a very small part of effective urban warfare, and even when required, it is better handled by troops with mobile, protected firepower—the very definition of tank and mechanized infantry forces.

The other great misconception about urban fighting is that it is inherently disadvantageous to U. S. forces. Again, just the opposite is true. Although we must not underestimate the challenges associated with city fighting, it is important to see the many advantages as well. Urban environments will most often offer shelter, power, water, fuel, communications, and powered lift devices (for maintenance and supply operations), just to name a few. Further, cities permit ready access to the populace that all military operations must ultimately influence. Effective political penetration of the urban population can result in reinforced intelligence operations as a network of citizens supplements our other reconnaissance and surveillance instruments. Cities often have well-developed roads and other transportation means, such as airports and rail lines.

More to the point, our preparation for the future must include the fact that to categorize an environment merely as "urban" is a gross oversimplification. One can hardly compare the confusing, poorly marked dirt alleys of Mogadishu with the streets of Miami or Kiev. Instead of generalizing and then fearing the worst, we must analyze the challenges and potentials of the great diversity in urbanized terrain. And just as Field Marshal Slim came to perceive the fearsome jungles of Burma as opportunities for maneuver and victory rather than as obstacles, so we must train ourselves to see the powerful potential of American forces in urban terrain.

Thus far we have noted that the computer and sensor-related technologies are vital components of the Information Age. But just as the human brain and human senses are irrelevant unless connected by a nervous system, so the proliferation of computers and sensors is meaningless without communications. Modern warfighting does not depend simply upon *having* information, but rather upon *moving* it from place to place, from weapon system to weapon system. This is an important distinction, because to a large degree, the feasibility of Information Age warfare will be constrained by communications bandwidth and data-flow rate.

Communications technology is slowly catching up with visionary warfighting concepts. Almost as soon as digital information began to come into general use, the demands for ever-bigger communications pipes commenced. For the less technically inclined, imagine information as five o'clock traffic on a busy city street. The rate at which the traffic can progress is directly related to the width of the street and the number of lanes. In order to push a heavy volume of traffic through the city, drivers want wide roads. Likewise, in order to move information around the battlefield, soldiers require large communications pipes.

It may be difficult to predict the precise nature of the future battlefield, but we may be certain that it will be built on a complex array of digital and voice communications networks. Those networks will include and integrate both hard-wired and wireless links, both satellite and line-of-sight signals. Just as the infantry-based armies of the ancient world depended upon roads to move couriers and messages, so future armies will rely on redundant communications networks.

INFORMATION AGE WARFARE

Having defined the Information Age qualitatively, the next task is to determine the character and principles of Information Age warfare. Early doctrine on information warfare defined it as "actions taken to achieve information superiority by affecting adversary information, information-based processes, and information systems, while defending one's own." The past several years have witnessed myriad articles and books on Information Age warfare, but each author approaches the concept from a different angle. To some writers, Information Age warfare is all about the mass media and its relationship to military operations. To others, it is chiefly concerned with precision weaponry. Still others consider it to be centered on electronic warfare or psychological operations. The first step in assimilating these various interpretations is to enumerate the available visions of Information Age warfare.

Command and Control Warfare

To begin with, future warfare will be characterized by a dramatically increased emphasis upon "command and control warfare" or "C2W." In layman's terms, C2W is a dimension of conflict in which opposing armed forces attack each other's information systems and processes, while protecting their own. The intent is to create a condition in which the friendly side can perceive the battlefield, control its forces effectively, and act decisively, while the enemy cannot.

Probably the biggest roadblock to an effective information-warfare doctrine is that people (both civilian and military) who write about it have too narrow a viewpoint. Typically, articles and books about information warfare key in on specific technologies related to computers, sensors, or communications. But information warfare is a much broader concept than any of these components.

Information warfare includes political and social dimensions. It embraces psychological operations, media relations, and civil affairs. It employs not only destructive computer viruses and the jamming of communications, but also such age-old techniques as camouflage, feints, and ruses. Even a simple bombardment of an observation post or a command and control facility contributes to information warfare. Something so mundane as well-thought-out staff procedures that result in faster decision making are also part of the struggle for the information advantage.

The best way to avoid erroneous conclusions about information warfare is to remember that commanders have used it throughout history. Alexander, Caesar, Belisarius, Suleiman, and Genghis Khan would all have been comfortable with our doctrines about deception, civil affairs, and operational security. Information warfare has a long historical tradition behind it; only the components have changed.

The Media
Information Age warfare is characterized by its increasingly scrutinized relationship with the mass media. As Alan Campen noted in his book on Operation Desert Storm, the military used half as much bandwidth in prosecuting the Persian Gulf War as the press did in reporting it. But the truly revolutionary aspect of media coverage is the speed with which journalists can transmit details of military operations. Whereas in the past, the media could reveal only *recent* operations, in the Information Age, they can report *current* activities as well. If security concerns were a source of media-military friction in the past, this new capability only exacerbates the problem.

Someone once suggested that the basis for the conflict between the military and the media is a turf war: Both institutions claim to be the protector of [American] freedom. The media's relationship to personal and political freedom is indeed viable and constitutional. Since before the social contract of John Locke, Western thinkers have viewed the media as a critical component of political intercourse. The journalist is the common man's means of surveilling the government. Hence, in order for a republic to function properly, the media must have unrestricted access to government (including military) activities.

Enter electricity. Since the development of telegraph, telephone, radio, television, and satellite communications, the media have consistently closed the time gap between event and report. The acquittal of O. J. Simpson, as a dramatic example, was broadcast in real time across the world. The result of this accelerated reporting has been ever-increasing encroachments on military security. But of late, the time gap has virtually disappeared altogether. Further, satellite communications links allow the journalist to communicate directly to the home front, bypassing military authorities. There may well be a new constitutional issue on the horizon: Does the media have the right to report on current military operations? Alan Campen suggests that the answer is a decided negative, and he comes at the question from a philosophical rather than a military viewpoint. His point is that journalistic reporting is relevant to the function of democracy only when the people can influence the activity reported. Hence, the Framers' intent in protecting the media was to allow the people access to information that they could use in their interaction with the government. Since the people cannot (and should not) influence the prosecution of current tactical operations, they have no inherent right to knowing the details of them. Although they can and should influence the course of a *war*, they cannot and should not influence the course of a *battle*. Further, since enemy agents can make use of such current information, the media should be excluded from such reporting.

All of this suggests that when viewed conceptually, the media no longer simply *report* military activity, but rather *participate* in it. Strategy, operational art, and tactics must now accommodate the presence and effects—both positive and deleterious—of media operations. But if this is the case, then a fundamental and even constitutional question emerges: Should American military art deliberately embrace the media as a tool in warfare? On the one hand, cultural sensitivities and a large body of public law would suggest that deliberate use and manipulation of the media on the part of military officials would be wrong and perhaps illegal. On the other hand, strategic military theory would condemn ideologically based notions of a strictly lais-sez-faire approach to media communications as romantic nonsense. Ultimately, history will scoff at such eighteenth-century pretensions just as today we laugh at feudal prejudices against gunpowder. But in the end, these two points of view will have to reach a compromise so-lution if the United States is to prosecute military operations effec-tively and at the same time retain its identity as a free nation. Further, since both military and media operations fall somewhere within the social sciences, we can expect that any attempt by one to influence the other will cause a series of reactions and counterreactions until a sort of equilibrium is reached. One thing is certain: the relation-ship between the modern military and the modern mass media can-not be governed by a philosophy developed in the age of the print-ing press and telegraph.

Precision Strike
Of all the components that compose Information-Age warfare, the capability of precision-strike operations is probably the most publi-cized and recognized. The advent and rapid advances of smart bombs and their related equipment have conferred a unique char-acter on United States armed forces as the vanguard of Information-Age warfare. Compared to the seemingly ham-handed mass bomber attacks on Germany in World War II, the one-sided decapitation of Iraq's military capacity during Operation Desert Storm appeared as a model of the new warfare techniques.

Since the invention of missile weapons in ancient times, there has been a problem with precision targeting. The circular probable er-ror (commonly referred to as 'CEP'), as a modern artilleryman would say, was a constant factor in the planning and employment of missile weapons, from the sling to the howitzer. CEP can most eas-ily be understood as a circle in which the center point is the aim point of a missile weapon. The more inaccurate that weapon, the larger the circle in which the round might actually fall. The logic of the CEP is simple: the larger the CEP, the more rounds are required to achieve the requisite damage on the target. The converse is also true: the smaller the CEP, the fewer rounds are required.

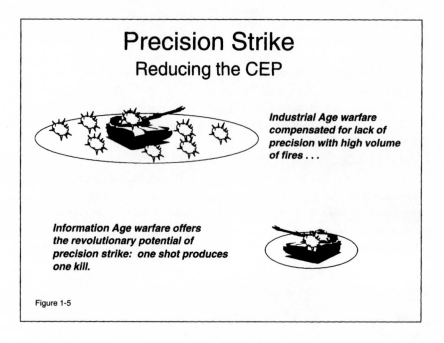

Precision Strike
Reducing the CEP

Industrial Age warfare compensated for lack of precision with high volume of fires . . .

Information Age warfare offers the revolutionary potential of precision strike: one shot produces one kill.

Figure 1-5

Industrial Age technology addressed the CEP problem with focus, determination, and skill. The Norden bombsight, infrared sensors, the science of ballistics, to name a few innovations, chipped away at error significantly to the point that artillery and other long-range fires grew to be the most lethal weapons in war. But even at the end of the Industrial Age, CEP still existed and was a factor in military art and science.

The Information Age offers new potential. Through the fusing of computer technology, global positioning system (GPS), advanced sensors, digital mapping, and sophisticated communications networks, weapon systems can achieve the Information Age phenomenon of precision strike, eliminating—theoretically at least—the problem of CEP. "One round–one kill" (or even "one round–multiple kills") can become a reality in the modern age. Precision strike, a hot topic in contemporary military writings, is a real capability. What remains is the question of whether the military establishment that achieves precision strike can adjust its military theory, principles, and tenets in order to exploit the possibilities offered by the new technology. And even more important, can the military establishment avoid the temptation to cling to precision strike as its sole approach to warfare? Overreliance upon a single capability in war guarantees failure and defeat, as we shall explore further in this book.

In a broader sense, precision-strike capability gives rise to a philosophical question concerning modern war. If a nation were able to bring precision strike to a high state of maturation, how would this capability impact on the acceptability of war to the populace? Suppose, for the sake of argument, that a nation's armed forces could develop a precise ability to strike at exactly the right enemy with the correct weapon at the most effective point in time and space, with no collateral damage and little risk to its own armed forces. Would such "clean" warfighting capability increase the population's willingness to accept warfare? If so, then one of the implications of precision-strike capability might well be a general increase in the citizens' affinity for war.

Precision Movement

Hand in hand with advances in strike capability, the U.S. armed forces are also advancing, though less spectacularly, in the ability to move with precision. Precision movement, in fact, pervades today's armed forces at all levels of war.

The problem of military movement throughout history has had both logistical and tactical dimensions. The logistical function of transportation has been at the forefront of notable disasters down through the ages. Incorrect loading, storms at sea, air disasters, lost or misdirected supplies, and other mishaps have characterized military movements in the past. Tactically, military units have been lost, out of position, in the wrong formation, and otherwise dislocated since the time of Sargon of Akkad. Even in recent times, tactical leaders have spent a great deal of time searching for their subordinates. But the Information Age has introduced technology that significantly impacts on these problems.

The computer has aided movement planning and execution to an enormous degree. Today's army is a "force-projection army." That is, rather than relying on permanently stationed troops all over the world, the U.S. Army is primarily based in the continental United States and will, in time of need, project its fighting forces abroad through air and sea lift. Without doubt, such a bold claim could not be realized were it not for the myriad computer systems and software that automate the planning and execution of large-scale movements. In short, without computers, the U. S. armed forces could not move.

Global Positioning System (GPS) has also changed the nature of military movements. GPS is a system of satellites in geosynchronous orbit around the earth, each of which broadcasts signals to radio receivers on the ground. The receivers interpret the signals and convert them into precise positioning data. By simply reading a digital display, the soldier can know where he is on the ground. One Army colonel, after coming to grips with how GPS assists in movements, exclaimed, "We've been lost all these years and didn't even know it!" The colonel's finding is exaggerated but essentially accurate. Prior to GPS, a tactical leader had to navigate by using terrain association, a map, and a compass. With such a system, disorientation in movement was as common as CEP in artillery fires. The savvy officer planned for some degree of disorientation by allowing time for reconnaissance, posting guides, and rehearsing movements.

GPS provides a true leap-ahead capability in moving. The first order effect of the technology is that a military leader can know precisely where he is. By overlaying a communications network on the military unit, the leader can know, in the second order, where his subordinates and superiors are. Finally, by adding yet another component, such as a laser rangefinder, and a computer to integrate the systems, the soldier can shoot the directional laser at the enemy and learn precisely where the enemy is as well.

Another component of precision movement is Total Asset Visibility, an ongoing program in the American armed forces. Through a system of signaling devices placed on or in military containers, robust communications networks, and associated databases, military and civilian shippers can identify, track, and accurately route supplies and equipment from the base to the field unit. In essence, each pallet or container of supplies periodically "announces" its whereabouts, contents, and condition to the system using radio communications. Every time the container is moved or its status otherwise changes, the system can be updated. Further, transportation personnel on the ground can identify the contents of unopened packages. This new capability leads to accurate, efficient movement of supplies and equipment. Again, the theoretical and philosophical implications of such capabilities have perhaps not been thoroughly explored. Since logistics is the soul of operational movement, these advancements impact directly upon operational tempo.

Transportation means have likewise advanced in technology. The C-17 air transport provides a key capability to a military force that bases its strategy on rapid and accurate movement. By being able to land on smaller, less-prepared runways, the C-17 provides many more options for entry into a theater of war. As a result, the force can move more precisely and rapidly to where it must operate.

Roll-on, roll-off (RORO) ships and fast sea lift likewise aid precision and rapid movement. By allowing military units to board the ship and exit the ship in the same order of march, RORO shipping reduces the complexity of planning and provides greater flexibility in overseas movements. Further, self-contained cargo ships (i.e., cargo ships that have onboard lifting and loading devices) enhance precision movements in the same way as a C-17 enhances air movement: by providing more numerous options for off-loading. Finally, sea-transport technology is advancing the speed of seaborne deployment to a previously undreamed-of rate of movement. When coupled with the current practice of prepositioning wartime stockages of equipment and supplies near threatened theaters, the potential leap in strategic agility is awesome.

In a more holistic sense, precision movement includes precision mobilization. In this regard, the U.S. military has improved dramatically since the days of World War II. As the vital counterpart to the active force, America's reserves have become fully integrated not only into the total armed force, but more importantly, into America's strategic formulation process. America cannot even contemplate significant military action without some degree of reserve mobilization. In the place of the recently demised Selective Service, which was a mobilization system geared to the production of Industrial Age mass armies, the United States has opted for "precision mobilization" (to coin a term). Today, the reserves are the reposi-tory for selected, highly specialized skills, primarily combat-support and service-support skills, without which America's armed forces can-not move or operate. Hence, a vital component of precision move-ment is precision mobilization.

Precision Protection
The other major area in which the American armed forces are de-veloping precision is force protection. The best way to think about the protection issue is to consider alternative methods for protect-ing a given weapon system. One way is to surround the system with armor plating. This method has been prominent at various times throughout history. Although armor can indeed provide reliable pro-tection, it is often costly in terms of money, weight, and mobility.

The other method of obtaining protection orients on the attack-ing system. Rather than increasing the capacity to absorb attack, the second method looks to defeating the system that produces an at-tack. In a broad sense, the old proverb "offense is the best defense" is an expression of this method. But in the context of the Informa-tion Age, the systemic approach to protection is more complex. In essence, the systemic approach begins with an analysis of the enemy's attack procedure. It asks the question: What must the enemy do in order to produce an attack against the friendly weapon system? Given the complexity of modern weaponry, the answer to that question will probably reveal a number of discrete steps that the enemy must per-form sequentially in order for his system to function effectively. For example, for an enemy missile to destroy a friendly weapon or in-stallation, the enemy must begin by procuring (through production or commerce) both the missile and its launcher. He must mobilize and train a crew. He must deploy the weapon to a position from which it can hit the target. He must employ an intelligence system or systems that can acquire and track the target. Further, he must develop a command and control system that can translate political ambition into military orders. In summary, the missile attack requires myriad steps—logistical, administrative, and tactical—many of which are so critical to the outcome that, if interdicted, would nullify the attack or reduce its effect.

Having comprehended the enemy's attack system, the next step is to determine which elements of the system are both critical and vulnerable. It may be that the friendly force does not have the means to prevent production or procurement of the weapon system. Alternately, the friendly commander may be capable of interdicting fuel supplies, but the interdiction will not serve to disrupt the attack, because the enemy has already stocked fuel in forward areas. But it might be possible, for example, to deceive or confuse the enemy's sensors, thus preventing effective attack. Or perhaps the commander can interdict the enemy's ability to accurately assess the damage the attack inflicts, thus to some degree reducing the effects of the attack.

Precision protection uses this method in order to most effectively and economically prevent successful attacks on the friendly force. The Patriot missile is a good example of precision protection. During Operation Desert Storm, the Patriot successfully intercepted the incoming Iraqi Scud missiles. Rather than spending the enormous amounts of time and money that would be required to fortify friendly assets against Scud missiles (arguably an unfeasible proposition anyway), the American-led coalition used Patriots to knock down incoming missiles and other airborne assets to detect and attack the launchers.

Since the conclusion of the war, the effectiveness of the Patriot has come under scrutiny. It may well be that the Patriot's effectiveness was limited, but even this conclusion leads to another insight on precision protection. During the actual conduct of the war, the media's coverage of the Scud attacks and the Patriot defenses gave the *appearance* that the Patriot could effectively nullify Scud attacks. This appearance arguably was more important than the actual performance of the Patriot, because the media's reports left impressions on the leaders of Iraq, Israel, and the coalition states. In effect, the media, perhaps unwittingly, served as an agent of precision protection, robbing the Iraqi attacks of their potential political effects through the manipulation of information.

Another excellent example of precision protection is the recent development of "slew-to-cue" technology in ground-based air defense. Air-defense radars, computers, and communications are all old technologies, but we have now integrated them effectively to permit a leap-ahead capability to defend against enemy aircraft. With slew-to-cue technology, an air-defense radar can detect an inbound aircraft and, through digital communications, alert a firing battery miles away. The weapon system can then be automatically oriented toward the approaching threat, long before visual detection occurs. As a result, the air defender can engage and destroy the aircraft at the maximum range of his weapon. The information advantage makes the air-defense missile infinitely more effective.

Precision protection will continue to be a theme in American warfighting. The continued exploitation of the electromagnetic spectrum, from gamma radiation through radio waves, will enhance our ability to pick apart the enemy's attack system and interdict the vulnerable and critical links. Combined with the less-selective protection offered by armor and other "attack-absorption" means, precision protection will offer an Information Age army greater ability to protect the force.

Psychological Operations

War is a psychological phenomenon. Whether proceeding from a Clausewitzian viewpoint that war is an extension of politics, or from the point of view offered by John Keegan that cultural factors are the determinants in war, the student of history quickly appreciates that warfare emanates from the human soul. Operations intended to impact directly on the mind and willpower of the enemy have always attended armed conflict. Ancient armies sometimes slaughtered whole city populations that resisted siege while treating those cities that surrendered more leniently. The purpose of this policy was to convince potential enemies not to resist. Military uniforms and equipment have been designed at various times in history to frighten and discomfit the opposing side. More recently, loudspeaker teams and air-dropped leaflets have been used to demoralize the enemy in battle.

Efforts at psychological manipulation will certainly continue and expand in the Information Age. It is traditional within the U.S. Army to analyze a potential threat in terms of both capabilities and intentions. Psychological warfare takes aim at the latter. By targeting civilian populations, soldiers, military leaders, and even political leaders, psychological operations attempt to interdict, disrupt, or destroy intentions. In broad terms the army employing psychological operations attempts to manipulate information in order to cause the enemy to believe either that they *cannot* oppose the friendly force, or that they *should* not. When a conflict is between two armies that are trying to influence the other psychologically, the result is a struggle for the informational high ground.

As the practice of psychological operations matures, PSYOP and maneuver will no doubt become more complementary than was true in the past. The former will attack enemy intentions while the latter will oppose enemy capabilities. In a well-designed campaign, the psychological attack on enemy intentions will eventually degrade enemy capabilities, and vice versa. The struggle for the informational high ground will not be the only focus of future operations, but it will no doubt be fundamental to them.

Civil Affairs
Another dimension of Information Age warfare that has ancient roots is civil affairs. The American armed forces have been conducting civil affairs for years, but recent doctrine has strengthened the tie between civil affairs and tactical operations. In the past, civil affairs were often viewed as "that nonmilitary stuff you do after a war." This perspective was an outgrowth of twentieth-century ideas of total war, which viewed war as composed of two unrelated phases: first, the total destruction of enemy capacity to fight; and second, compelled obedience of the defeated population. Although such a pattern of warfare pertained only to a very brief period of recent history, the assumptions that underlay it persisted long after total war was no longer a serious threat. As a result, American thinking on the complexities of civil affairs in a military scenario lagged behind other aspects of our doctrine. Today, our doctrine is advancing to an appreciation that an effective policy toward civilians in a theater of operations is a key to tactical success. Political preparation of and penetration into a theater of war should occur before, during, *and* after tactical operations.

Conclusion

In this chapter, we have established the characteristics of future warfare that are changing most dramatically. Before we continue into an examination of the existing principles of war, we must take a moment to consider the degree to which knowledge and ignorance dominate modern warfare. To do that, we will examine a well-known campaign from the past—but from the perspective of information flow. We will see how it was *truth* or the lack of it that decided the outcome. Armed with this vision of the future, we will then proceed to understanding what the principles of war are and why they need to change.

2: Maryland, 1862:
The Information Campaign

The commander . . . finds himself in a constant whirlpool
of false and true information . . .

—Clausewitz

The present seems to be the most propitious time since the
commencement of the war for the Confederate Army to enter Maryland.

—Gen. Robert E. Lee, September 1862

My people are destroyed for lack of knowledge.

—Hosea 4:6

Bloodshot eyes reflect the eerie glow of a computer screen, while
too much caffeine causes the quickly moving fingers to tremble
slightly as they poke out a message on the keyboard. Earlier in the
evening, a sudden system crash robbed the weary warrant officer of
a few hours' sleep, and the harsh cold and pressure of the battle
made recovery even more of a task. Even when aided by the Army's
latest experimental technologies, life in the field can be misery.
Now, at three in the morning, Information Age warfare doesn't
seem very glorious.

But it is.

After we sort through all the megahertz and kilobytes of the army's
futuristic experimentation, we will find that through the Herculean
efforts of many soldiers and civilians, the U.S. Army and her joint
partners have advanced the armed forces into a whole new age. Skep-
tical as we often can be of our own efforts and beliefs, we are without question standing at the dawn of a revolution in military affairs.

But it can be difficult, when immersed in the details of information technology and doctrines, to understand just how fundamental information really is to warfare. Further, as we continually anticipate enemy reactions to our own modernization, a clear vision of
future battle is an elusive target. As we strive with the contrived

enemy force, each side struggling for a minuscule edge in information, it is easy to lose perspective on how truth and ignorance can decide battles, campaigns, and wars. *How* will information be useful in the future? What effect will an information advantage really confer on the one who has it?

To illustrate the potential effects of information dominance, let us climb into a time machine and journey back to September 1862. We are standing on the south bank of the Potomac River, gazing into Maryland. Beside us, relegated to riding in an ambulance due to a recent injury, Robert E. Lee contemplates the next few weeks. The campaign for Maryland is about to begin. And it is a campaign that will be decided not by guns, rifled muskets, or cavalry, but by *information.*

Recently, I was privileged to participate in a tour of the 1862 campaign in western Maryland that ended with the Battle of Antietam (or Sharpsburg). It was a pleasant break from my routine duties, but as I toured the various routes of march and battle sites, I found to my surprise that the same themes that guide our efforts in modernizing the army were present 135 years ago. We are looking to state-of-the-art digital technology today to answer the same basic questions that commanders in the Civil War needed to resolve:

Where am I?

Where are my buddies?

Where is the enemy?

The Maryland campaign turned on information. The results of the campaign—some claim that it sealed the fate of the Confederacy—cannot be explained by the mathematics of numbers of muskets and cannon. Like many military campaigns before and since, the Maryland campaign was won and lost based on knowledge and ignorance. And because both commanders, McClellan and Lee, lacked clear perception of the area of operations, both sides' performance was characterized by failure.

In this chapter, we shall take a quick trip through western Maryland in September 1862, and note along the way how information or the lack of it was fundamental to the outcome.

LEE PONDERS INVASION

After Lee's army had smashed John Pope's army of Virginia at the Second Battle of Manassas, Lee estimated that the disorganized Federals would be unable to act decisively for at least two weeks. He developed a grandiose plan to invade Maryland during that interval, and the student of history today can, on the surface, appreciate the good generalship Lee displayed in pushing to exploit his recent victories. But did General Lee have a solid rationale for his invasion?

Lee fully expected the citizenry of Maryland to rise up in support of the Confederacy when his Army of Northern Virginia moved north. Perhaps they would even secede, thus isolating Washington, the Federal capital. The decision to invade or not should have been made by Jefferson Davis, and the two men had certainly discussed the idea. But General Lee's sudden invasion, a few days after the victory at Manassas, preempted the political decision-making and committed the Confederacy to an invasion of the North.

In the end, Marylanders chose not to rise up against the Federal government. Lee's plan was fallacious, and the primary reason for invasion was invalid. If Lee had known that the citizens would not support him, would he have committed to such a desperate enterprise? No doubt, the course of the campaign would have been significantly altered at least. But why didn't General Lee know the truth?

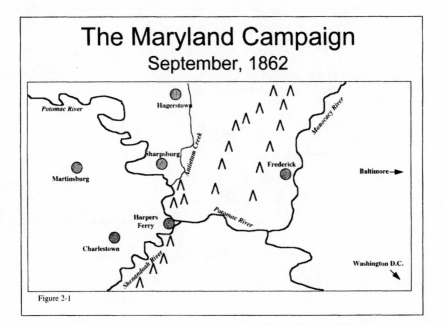

The Maryland Campaign
September, 1862

Figure 2-1

The Army of Northern Virginia was not organized for information. Lee's staff was pitifully small and totally inadequate for conducting modern warfare. If he had been served by a professional intelligence staff, they might have informed the general that most of the Confederate sympathizers were in *eastern* Maryland, not in the west, where Lee intended to march. Since there were comparatively few slaveholders in the west, the Confederate high command might have anticipated a less-than-enthusiastic response as his troops marched north. In the end, Lee was misinformed and ignorant of the true political situation. The campaign was imperiled by ignorance from the start.

SURPRISE AT FREDERICK

After an initial march to Frederick, Maryland, Lee turned west. A false rumor about a Federal force marching south from Pennsylvania toward Hagerstown caused Lee to send Longstreet north to intercept, while Jackson concentrated on taking Harpers Ferry. Mean-

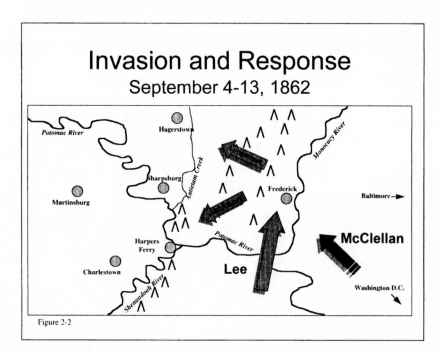

Figure 2-2

while, Daniel Harvey Hill was given the task to remain near South Mountain. Little did Lee or his generals realize how hard pressed Hill was soon to be. Lee was unaware that McClellan had departed from Washington in pursuit, nor did he know how fast the Federals were closing. As a result, Lee dangerously dispersed his army and invited destruction in detail.

Why did he imperil his command in this way? What might have happened in the Maryland campaign if Lee had had accurate and timely knowledge concerning the movement of McClellan's army?

Today, theater, corps, and division commanders (and even brigade commanders) have access to a variety of sensors that aid them in avoiding the disastrous ignorance that Lee suffered from. If we overlay today's technology on this campaign, General Lee would have detected the mechanized Army of the Potomac as it moved west. Joint Surveillance Target Acquisition Radar System (J-STARS), satellite imagery, airborne electronic intelligence (ELINT), and unmanned air vehicles (UAVs) would have warned Lee that a major move was under way.

If Lee had known about the Federal movements, what might have happened differently? Without a doubt, he would have concentrated for battle sooner than he did. Further, I suspect that rather than move westward, Lee would have defended along the Monocacy River, rather than at Sharpsburg, so as to give himself some room to maneuver and, if necessary, fall back. Instead, he fought with his back against the "wall" of the Potomac at Sharpsburg, and if he had faced a more capable general, the Army of Northern Virginia might have faced utter destruction. Once again, it was a lack of truth that precipitated a near disaster.

THIS IS TOO GOOD TO BE TRUE!

What would you do if someone suddenly handed you a complete picture of the enemy situation in war? Would you act upon it, or would you remain cautious, suspicious, and timid?

On 12 September 1862, a copy of General Lee's Special Order 191 fell into McClellan's hands. It detailed the movements and intent of the Confederate army and revealed the precarious situation that Lee was in. General George B. McClellan held in his hands a golden opportunity . . . and did nothing with it. At first, he believed it was a ruse. By the time he acted upon the information, the Rebels had concentrated again.

As I considered this infamous failure, I was struck by how it paralleled events in recently concluded Advanced Warfighting Experiments (AWEs) in the army. When information technologies suddenly delivered remarkably clear and accurate reports concerning the enemy, experimental Blue Force commanders were hesitant in using the information. Having served previously with the men involved, I know them to be bold and imaginative leaders, so what was the problem?

Clearly, there must be a strong leader development effort to accompany the technological advances we are making. Today's tactical leaders are not timid, but they have been trained within the context of the utter ignorance which has characterized warfare in our age. When technology hands us a clear picture of battlefield truth, it strikes at the heart of our doctrine, organization, and tactical concepts. Our battlefield formations, planning procedures, and tempo are founded upon ignorance of the battlefield, which heretofore has been fundamental to warfare. If we get to the point at which leaders can reliably, accurately, and instantly see the truth on the battlefield, our methods and practices must radically change. In the meantime, we may reflect on the accusation that actor Jack Nicholson yelled in a recent movie: "You can't *handle* the truth!"

AT CROSS-PURPOSES IN THE MOUNTAINS

Technology cannot compensate for indecision. In the last hours before D. H. Hill's tiny division faced an attack by two Federal corps at South Mountain, he was still uncertain as to what his mission was. There was some confusion as to whether Hill was to remain at Boonsboro in order to cut off the retreat of Federals escaping from Harpers Ferry, or whether he was the rearguard intended to slow the advance of the Army of the Potomac westward. Was Hill supposed to work in coordination with Jackson and thus orient south, or was he working for Longstreet and thus to face east against McClellan?

In the end, the latter mission was thrust upon him, and Lee could thank Hill's tenacity and John Gordon's ferocity for saving the day at South Mountain. But as I stood in Turner's Gap and considered the bloody task that Hill was handed on 14 September, it was clear that even perfect information cannot replace a flawed concept of operation. Even when served with the latest technology, commanders must be clear when issuing their orders and intent.

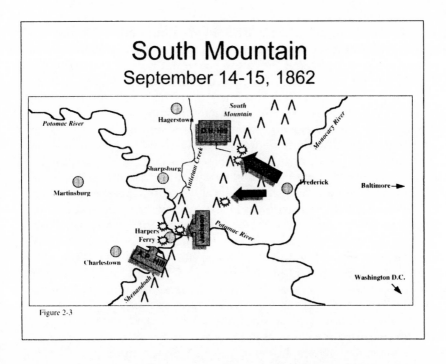

South Mountain
September 14-15, 1862

Figure 2-3

SITUATION AWARENESS AND VELOCITY

Information makes us move faster. At least, the *potential* for more rapid movement is there. It remains to be seen whether we will take advantage of the velocity advantage that information can confer.

Why should information affect movement rates? Facts cannot increase the physical speed of tanks and infantry fighting vehicles, much less foot soldiers, so how can they make us move faster? The answer is that information changes not *vehicle* speed, but rather *unit* speed. An armor battalion does not move at an average rate of sixty kph, even if its tanks are capable of that speed. Rather, the battalion moves at a shockingly slow rate of twenty, fifteen, or even five kilometers per hour, depending upon terrain and weather. The major determinant in unit speed is not weapon-system speed, but rather the unit's formation. Units move quickly when in column formation, considerably slower when in wedge or diamond formation, and they are almost stopped when in line formation. Hence, to increase unit velocity, change the formation.

Of course, units on the battlefield are not often in column formation, because of the danger of unforeseen contact with the enemy. In the past, even if the enemy were not nearby, it was the *possibility* of enemy presence that forced commanders to change from fast-movement formations into formations that traded off speed for protection and firepower. But what would happen to our relative velocity if we had a clear understanding of where the enemy was, or more importantly, where he was *not?* If we could be certain that the enemy could not oppose us in selected portions of the battlefield, then we could dispense with formations that trade off speed for security, and instead move at considerably greater speed in columns.

The other way that truth increases velocity is by informing our movements *toward objectives that count.* When the commander perceives the area of operations accurately, he does not waste time and energy moving hither and yon toward things that don't matter. He doesn't chase rumors, guard against attacks that never come, or search for something that isn't there. Instead, the movement of his troops becomes intensely *purposeful*—every step of every soldier advances the friendly plan. A comprehensive military history of the past 4,000 years might well include a chapter on irrelevant marches. We have arrived at an age, however, in which that chapter could come to a close. Velocity is a vector quantity, a straight-line phenomenon. Truth makes our paths straighter, and thereby increases velocity.

An increase in tactical and operational velocity is a potentially revolutionary advantage. Faster movement means hitting the enemy before he is ready, giving rise to more opportunity for surprise, rather than attrition fights. But this advantage will materialize only when we convert information into velocity, and that conversion is an act of volition and determination on the part of the commander.

Likewise, tactical and operational velocity will benefit from situational awareness—a term which describes a condition in which the commander knows the answers to the three central questions:

Where am I?

Where are my buddies?

Where is the enemy?

By answering these three questions, the friendly force can not only manipulate their formations as described previously, but they can also appreciate the *need* for rapid movements when those opportunities arise.

Such an urgent opportunity came about in mid-September 1862, as McClellan slowly advanced westward. A crisis was occurring at Harpers Ferry, but McClellan did not move rapidly enough to save the day. Unaware of the desperate situation faced by the Federal garrison there, McClellan allowed his troops to plod along while a key installation fell to Jackson and McLaws.

But what might have happened if McClellan had been served by situational-awareness technology? Suppose, for a moment, that he had been able to see the entire situation within Maryland on the evening of 13 September. We may well believe that he would have stirred his ample forces to exploit the halfhearted Federal breakthrough at Crampton's Gap—a breakthrough that threatened to destroy McLaws's force, which was at that moment sealing the fate of Harpers Ferry. Instead, awash in ignorance of developments a scarce twenty miles from his headquarters, McClellan permitted his army to crawl forward, missing opportunity after opportunity.

DIGITAL COURAGE AT THE PRY HOUSE

McClellan set up his headquarters in the Pry House overlooking Antietam Creek as Lee's army sprinted to concentrate near Sharpsburg. History records for us the maddening inertia of General McClellan, and even his apologists must admit to a certain timidity in "Little Mac" as he let slip a grand opportunity to strike a war-winning blow at Sharpsburg.

On 15 September, McClellan's forces outnumbered Lee by over two to one. Neither Jackson nor A. P. Hill had yet rejoined the Army of Northern Virginia, and there is little doubt that a vigorous attack on the fifteenth would have doomed the Confederates. But McClellan chose not to hazard an attack until he felt all was ready. That moment did not come on either the fifteenth or the sixteenth of September. It was only on 17 September that McClellan finally bestirred himself and attacked, by which time Lee's forces had more than doubled with the arrival of Jackson, and A. P. Hill was rapidly closing.

Why did McClellan fail to attack earlier? Part of the reason was that he lacked information. Technology cannot change a man's character, but it might have provided a bit of "digital courage" at the Pry House. For George B. McClellan believed that he faced close to 100,000 Confederate troops, when in fact he faced only 20,000 at first, and later about 45,000! This fivefold miscalculation of the enemy prevented the decisive victory the Union needed, and it led directly to McClellan's relief from command shortly thereafter.

Once again, the day was decided not by numbers, but by *knowledge* and *ignorance*. Lee, who began the invasion of Maryland with only some 55,000 men, had lost droves to desertion since the campaign began. Yet, in McClellan's mind, the Federals were outnumbered two to one!

ELECTRONS, BLOOD, AND KEY TERRAIN

The seventeenth of September saw the single bloodiest one-day battle of the entire Civil War. Thousands were to die on that day, but as I strolled from Dunker Church through the West Woods and glared at the famous Cornfield, I wondered how many lives might have been spared had Gen. Joseph Hooker had an accurate electronic map on a digital display.

Hooker's I Corps was responsible for conducting the northern envelopment of the Confederate lines. With near-perfect hindsight, we

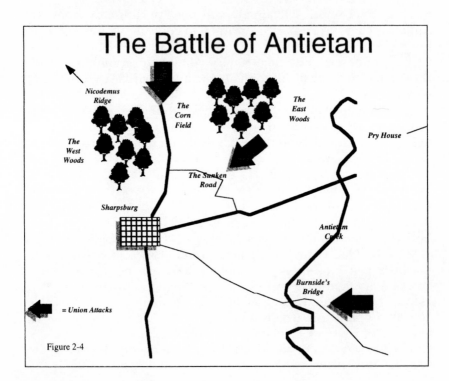

Figure 2-4

can confidently conclude that had "Fighting Joe" Hooker directed his corps more carefully, he would have collapsed the Confederate flank. Even with Hooker's (and later Mansfield's) bungling, Gen. John B. Hood was able to hold on through sheer guts and at the cost of nearly his entire division. Hooker's envelopment failed, but by inches and moments only.

Why did this attack fall short of the potentially grand success that lay before it? One critical reason is that Hooker neglected to seize a key piece of terrain on his right flank: Nicodemus Hill. On that commanding rise, Jeb Stuart's cavalry troopers guarded a Confederate gun emplacement from which a handful of artillery pieces tore at Hooker's attack. The moral and material effect of the Rebel artillery, when combined with the tenacity of Hood's defense, brought the Union I Corps to a halt. Lee's army was saved.

What might have happened if Joseph Hooker had had digital terrain? If he had planned and carried out his attack with greater precision, and with a fuller appreciation of the military effects of the terrain, would he have prevailed? Likely yes. Instead, his attack culminated, and another general—an engineer in charge of a column of mostly recruits—would have to continue the effort to crush Bobby Lee's left flank.

DESTROYED FOR LACK OF KNOWLEDGE

Joseph Mansfield, in command of the Union XII Corps, had been ordered to help Hooker. Feeling his way along unfamiliar roads on the morning of 17 September, Mansfield was uncertain as to what precisely he was to do. There was some confusion as to whether he was to relieve Hooker's corps or support it. As Hooker's troops bled and died, Mansfield's troops advanced slowly on their left.

Suddenly, General Mansfield heard firing ahead. His troops were shooting at something. Fearing that his soldiers were firing at Hooker's men, Mansfield galloped forward and commanded his men to cease fire. Moments later, he was struck by a Confederate bullet and died a little while later.

Where am I?

Where are my buddies?

Where is the enemy?

General Joseph Mansfield simply did not know the answer to those three questions, and he earned himself a monument that still stands at the spot of his last, disastrous ignorance.

Later that morning, General Sumner's II Corps advanced against the Confederate left. Through a trick of the terrain, his divisions split apart from one another—one heading toward the West Woods, and the other two toward the Confederate center. The resulting attack, the last desultory blow against Lee's left flank, fell apart after more bloody fighting.

There would be no decision on that side of the battlefield. Ignorance brought two armies to a crimson halt.

But . . . what might have happened if Hooker, Mansfield, and Sumner had had situational awareness? What if, as we have seen in our latest experiments, the understanding of unit positioning and movement had allowed three Federal corps to converge and synchronize their attacks? Almost certainly, the Army of Northern Virginia would have been annihilated at a stroke, and the war may well have ended soon after. In any case, the toll in blood on both sides would likely have been considerably less. Instead, the fight migrated to the center of Lee's lines—to a place known as the Sunken Road.

ORDERS AND DISORDERS

Daniel Harvey Hill had been posted in the center of the Confederate line. The early morning had passed with an ever-increasing din on Hill's left as Jackson's corps held on grimly. As previously mentioned, Charles Sumner's II Corps had split into two groups as it advanced. French's division closed in on Hill's position, and Richardson's group joined battle soon afterward. The Federal rifles drew blood from Gen. John B. Gordon's 6th Alabama, but the disaster that followed was not inflicted by gunfire.

When Gordon fell from a succession of five wounds, command of the Alabama regiment devolved on Lieutenant Colonel Lightfoot. Lightfoot, concerned that the Federals' enfilade was taking a toll on his men, sought out his brigade commander, General Rodes, and asked for permission to adjust his lines. Rodes assented. Lightfoot returned, shouting out orders over the din of battle. When a nearby major asked Lightfoot if his orders were for the entire brigade, Lightfoot erroneously said "yes." Suddenly, at the critical moment, Rodes's entire brigade turned and fled. The Confederate center had collapsed.

Why had this disaster happened? Did it result from the Federal volume of fires? No. Was it due to the opposing numbers of soldiers?

No. An imaginative and effective tactic on the part of the Union? Again, no. Rodes's brigade was routed by a lack of information—in this case, they misunderstood the commander's intent.

Fortunately for the Confederates, the Union troops paused and, with no effective resistance in front of them, they sought permission to exploit their success. Ultimately, McClellan refused, because to pursue this opportunity, he would have to commit Porter's V Corps— the army's last reserve. Because McClellan was unsure of the situation, he was unwilling to take such a risk. The Confederate center reformed.

As I stood on the spot in the Sunken Road where Gordon fell, I was again impressed with how information ruled the battlefield that day. Although information technology and internetted computers cannot retrieve a flawed plan, they can materially aid a commander in distributing combat orders and supervising their execution. Could Rodes's brigade have held if he had digitally burst a clear order to refuse the flank, instead of entrusting a lieutenant colonel with the message? If Rodes had been able to monitor the movements of his troops, could he have seen the mistaken retreat sooner and stopped it? What if McClellan had realized that there was no one between his soldiers and Sharpsburg as he hesitated near the center that day? But instead, a pervasive lack of understanding hung in the autumn air, suspended in a cloud of gunpowder, and blanketing the groaning wounded. The battle passed southward.

THE WIT AND WHISKEY OF AMBROSE BURNSIDE
The extreme right flank of the Confederate position had been steadily weakened by Lee as regiment followed regiment to the left and center. By early afternoon, a small contingent of 2,000 men remained to defend the stone bridge over the Antietam, known since that day as Burnside's Bridge.

General Ambrose E. Burnside began the battle in nominal command of a "wing" of the Army of the Potomac. His wing consisted, on paper, of the I and IX Corps. But when the fate of the battle came to rest upon him around noon, his wing had been divided, with the I Corps (under Hooker) deployed on the far right, leaving Burnside with only his old IX Corps on the Union left. Still, that corps could accomplish much if properly employed. With a vigorous and coordinated attack, Burnside might well have carried the bridge quickly and destroyed the weakened Confederate right. At the very least, he might have prevented reinforcements from opposing Union efforts elsewhere on the battlefield.

Instead, Burnside whiled away the hours, searching for an elusive ford across the creek and otherwise wasting time until his attack was too late to aid McClellan's other attacks. Then, when the attack came, it was so poorly coordinated and supported that the Confederates easily repulsed it several times. Finally, a pair of regiments stormed across the bridge—one was the 51st Pennsylvania, whose men had been cajoled into charging with a promise of a whiskey ration—and the creek was crossed.

Even then, Burnside's men dawdled, and did not carry the ridge overlooking the creek until two full hours passed since the taking of the bridge. By the time the attack against the Confederate right got fully under way, Gen. A. P. Hill's Light Division had arrived from Harpers Ferry, just in time to outflank the Union advance and bring it to a halt. Once again, Lee's army was saved in the nick of time.

Could a timely flow of information have changed the course of this final act of the Battle of Antietam? Without a doubt, a better knowledge of the terrain and situation would have strengthened Burnside's efforts. Had McClellan been able to fuse together data from situational awareness technology, J-STARS, and UAVs, could he and his staff have synchronized the Federal attacks to greater effect? We know that technology cannot reform the will and ability of a military leader, but it is inconceivable that the Army of the Potomac wouldn't have delivered a better showing that day if their generals had known:

Where am I?

Where are my buddies?

Where is the enemy?

As for the Army of Northern Virginia, we may well conclude that information would have precluded the near disaster of Sharpsburg altogether. Had Lee been better informed, we might instead be studying the Battle of the Monocacy, or, indeed, there may have been no Maryland campaign at all.

Conclusion

Information technology is not a panacea, and no one working the army's futuristic programs thinks otherwise. Information flow can *create* problems as well as solve them if we are not careful. But there is no doubt that the army is going in the right direction by thinking about, engineering, and fielding a digital force. Information, truth, and ignorance rule the battlefield and always have. In the past, we have had little choice but to acquiesce to the fundamental fog of war. But the twenty-first century holds out a new and exciting possibility: that future leaders can harness information to their advantage. In our experiments thus far, we have found that hypothesis to be shockingly vindicated. There are myriad challenges in building a digital force, but the potential for a true battlefield revolution is clear.

The bloody mathematics of the Maryland campaign fall short of explaining this remarkable September adventure. Tables of organization, numbers and types of ordnance, road distances, and logistical calculations combined simply don't tell the tale to the satisfaction of a student of history. The drama that played out over western Maryland in the autumn of 1862 was directed by knowledge, perceptions, false beliefs, sudden understanding, and a pervasive ignorance of what was unfolding nearby. Neither Lee nor McClellan needed reinforcements, supplies, or transportation as much as they needed the answers to three simple questions:

Where am I?

Where are my buddies?

Where is the enemy?

As we move toward the conclusion of this book, we will revisit the role of information on the battlefield. Knowledge and ignorance will together compose the central principle of war in the set of revised principles for the future. We are now ready to take on the existing principles of war. In Part 2, we will examine each one and see how information can shed much-needed light on some old ideas.

PART 2
The Principles of War

Foolish toy! Babies' plaything of haughty admirals, and commodores, and captains; the world brags of thee, of thy cunning and might; but what after all canst thou do, but tell the poor, pitiful point, where thou thyself happenest to be on this wide planet . . . no! Not one jot more! Thou canst not tell where one drop of water or one grain of sand will be tomorrow noon . . . Curse thee, thou vain toy . . . Curse thee, thou quadrant! No longer will I guide my earthly way by thee; the level ship's compass, and the level dead-reckoning, by log and by line; <u>these</u> shall conduct me, and show me my place on the sea. Ay, thus I trample on thee, thou paltry thing that feebly pointest on high; thus I split and destroy thee!

—Captain Ahab, <u>Moby Dick</u>, Herman Melville

3: Maneuver

And David came to Baal-perazim, and David smote them
there, and said, The Lord hath broken forth upon mine enemies before me, as the breach of waters.

—2 Samuel 5:20

Now an army may be likened to water, for just as flowing
water avoids the heights and hastens to the lowlands, so
an army avoids strength and strikes weakness.

—Sun-tzu

Maneuver: Place the enemy in a position of disadvantage through
the flexible application of combat power.

We begin with the principle of maneuver, one of the principles that
is in need of revision. The need for change is twofold: First, this principle needs to be reexamined because it has always been misleading
and inaccurate; second, the need for revision has become urgent in
the light of recent changes in warfare.

The problem with the principle as stated is that it is described as
both a *means* and an *end.*

> Maneuver is the movement of forces in relation to the en-
> emy to gain positional advantage. Effective maneuver keeps
> the enemy off balance and protects the force. It is used to ex-
> ploit successes, to preserve freedom of action, and to reduce
> vulnerability. It continually poses new problems for the enemy
> by rendering his actions ineffective, eventually leading to de-
> feat. (FM 100–5, 1993)

This statement describes physical maneuver (the means) as the
way to achieve an advantage (the end), with emphasis upon the for-

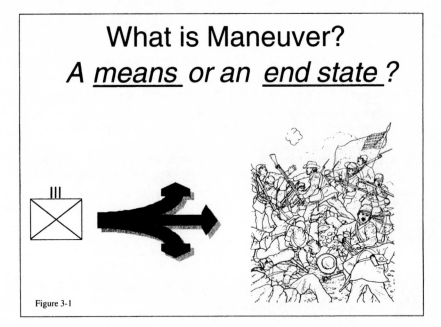

Figure 3-1

mer. Even the one-word name of the principle emphasizes the means to the end, rather than the end itself. It is analogous to creating a principle of economics entitled "Municipal Bonds." Rather than focusing on the need for secure investment, hedging against inflation, or maximizing returns, all aimed ultimately at increasing our wealth, we instead select a single method—investing in municipal bonds—and try to convince ourselves and others that this is the primary way to get rich. In the same way, the classic principle of maneuver describes the goal we are after—a disadvantaged enemy—but describes physical maneuver as the way to achieve it. This is incomplete and inaccurate. In order to advance our understanding of modern warfare, it is incumbent upon us to embrace maneuver as the *end state* of our plan, but not as the *means.*

The history of human warfare is a saga of continuous attempts to gain the advantage over the foe in battle. Before striking his murderous blow against his brother, Cain brought Abel into the field—where the victim would be isolated from the protection of the parents. Since then, warfare throughout history has featured a relentless drive to gain the advantage over the enemy before exchanging blows. While some cultural or religious ideas have glorified and insisted upon "fairness," particularly in single combats, most instances of conflict through the ages have featured each side struggling to obtain an advantage over the enemy.

If the desire for an advantage in combat is universal and intuitive, then where did the principle of "maneuver" come in? Is the physical maneuver of troops the only way to accomplish the goals of putting the enemy at a disadvantage, preserving freedom of action, keeping the enemy off balance, and so forth? Do I have to attack my foe from the rear or flank in order to procure an advantage for myself? The answer is a resounding "No!" and it always has been. In the practice of warfare through the years, there have been many methods and means to achieve an advantage.

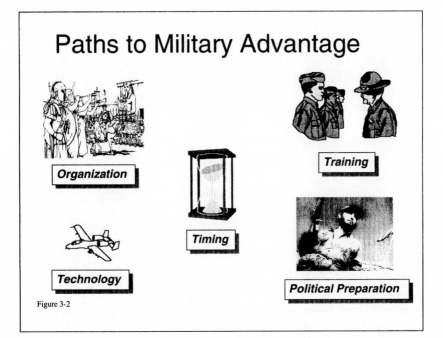

Paths to Military Advantage

Organization

Training

Timing

Technology

Political Preparation

Figure 3-2

Technology, for example, has often been used to gain leverage over the opponent in war. By the time of the reigns of Ashurnasirpal II and Shalmaneser III, the Assyrians had developed armies equipped with iron weapons, which gave them a significant edge (no pun intended) over enemies still using bronze weaponry. Even without any clever ground maneuver, the armies of Hannibal and his contemporaries were able to employ African elephants—the heavy battle tank of the ancient world—in battle to break Roman legions. And at the Battle of Königgrätz (or Sadowa), in 1866, the Prussians' technologically superior breech-loading rifle was able to retrieve their poor tactical coordination and win the day against the Austrians, whose infantry were armed with unwieldy muzzleloaders.

One of the themes of the twenty-first century will be the rapid development of technology and its equally rapid adaptation to warfare. Where mass production was the sine qua non of Industrial Age warfare, accelerated development and application of prototype technology will be one of the keys to success in the future. Hence, any principle of warfare that addresses gaining an advantage over the en-

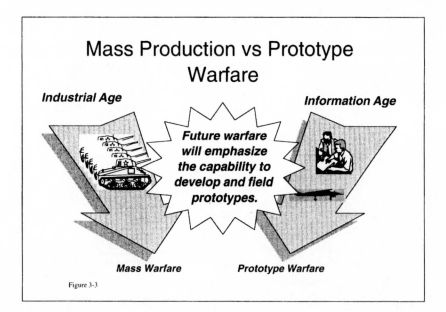

Figure 3-3

emy in the twenty-first century must speak to the intimate connection between armed forces and technological development. The principle of maneuver, as we have it today, does not address this.

Organizational innovations, such as the legion, the *corps d'armée,* and the panzer division have resulted in disadvantaged enemies. In each of these developments, the advantages of combined arms, efficient communications, and increased mobility foreordained battlefield success (temporarily, at least), quite apart from the actual maneuvers employed in battle. A modern-day business executive facing a slump in sales could try to analyze the problem and devise a clever solution. But a better approach would be to organize a team of engineers, marketing specialists, public-affairs personnel, and midlevel managers to create sustained success. A wise leader, rather than trying to solve the problem himself, will organize a team with the right components, confident that whatever solution they come up with will be effective, because the right people are working on the problem. In a similar manner, the effective organization and administration of an army can produce battlefield success, even if its tactical maneuvers are not optimal. Here the principle of maneuver also falls short.

Classical ideas about maneuver do not provide explicit guidance concerning temporal aspects of conflict. In the twenty-first century, commanders will have to be adept at manipulating time factors in their favor. We will look into temporal warfighting further when we take on the principle of surprise, but as we begin to change the old principle of maneuver to make it more inclusive, we must note that a critical part of disadvantaging the enemy is to alter the timing of operations. The four temporal characteristics that the commander can affect are *duration, frequency, sequence,* and *opportunity.* As I have described in a previous book, *Fighting by Minutes: Time and the Art of War,* soldiers can manipulate these factors in order to undo an enemy's plan. Again, the principle of maneuver, as we have it today, does not address the temporal means of gaining the advantage.

Moral and political preparation of the friendly force have often conferred the decisive combat advantage over the enemy. Much of the success enjoyed by the armies of the Roman republic or the early Islamic expansion came about because of the superior morale of the warriors. Likewise, in the European wars resulting from the French Revolution, the moral and political cohesion of French armies, once harnessed and disciplined by the war minister Lazare Carnot, overwhelmed their professional adversaries, as at Hondschoote, Menin, and Wattignies.

Modern theorists have noted that there is a merging of military considerations with political, economic, and cultural factors in modern warfighting to an unprecedented degree. The American generals of World War II may have been relatively free to romp across Europe in their drive to end Nazi Germany. They may have had the convenience of dividing activity into two categories: "military" activities and "nonmilitary" activities—and to concern themselves primarily with the former. But today, communications technology and the liberalization of political thought have combined to increase sensitivity to the political geography of any future theater of war. The modern general must think seriously about many more factors than fire and maneuver. Political, economic, and cultural elements exist not only as constraints, *but as positive opportunities to gain the advantage in conflict.* The grand strategy of the Roman Empire included political and economic penetration into conquered lands, as a hedge against possible rebellion and invasion. In Vietnam, the communists' strategy of *dau tranh* was essentially a political strategy, in which strictly military affairs were secondary. And while these aspects of war have always been present in history, they are much more fundamental to modern warfighting today.

The role of training in order to produce advantage in battle is also understated in the principle of maneuver. This is an odd omission, because a clear and oft-repeated lesson from the past is that trained soldiers possess a marked advantage over untrained enemies. Armies that are well equipped and trained have fought and won when outnumbered ten to one by lesser-prepared opponents. It is important that we understand the vital link between training and battlefield success, because current training strategies and resources will be strained to the maximum by rapidly developing technology. The most reliable weapon that the United States will ever employ is a disciplined and skilled soldier. When plans and preparations collapse and everything goes to hell during a battle, the trained fighting man is the best hope for maintaining the advantage over the enemy. The principle of maneuver, with its fixation upon the movement of troops as the means of gaining advantage, does not properly instruct tomorrow's warrior concerning training or other ways of getting the edge over the enemy.

To summarize the point, the soldier today has a great variety of tools at his disposal for putting the enemy at a disadvantage. The military implements of land, air, and sea warfare have been joined with technology that has opened the electromagnetic spectrum as a battlefield. Precision fires offer an ever-increasing opportunity to today's combatants. The integration of interagency capabilities, along with nongovernmental organizations, also gives military officials a new tool of conflict. Even the media, with its increased reach and effectiveness, factors into the drive for the advantage.

The physical maneuver of troops on the battlefield is just one method among many to achieve the advantage. The *goal* of obtaining an advantage over the enemy can be served by various *means*. The means will change from war to war, or even from day to day, but the underlying concept of *advantage* remains. It is this immutable idea of advantage that should be codified as a principle, not the transitory ways to get the advantage.

Then why has modern military thinking fixed upon maneuver as the chief means to the end? The reason is that the principles of war, as we have them today, were fully developed in the Industrial Age and in Europe. Within the context of warfare from Napoleon to World War II, physical maneuver truly was the predominant method for gaining the advantage in combat. The apex of Industrial Age warfare would typically pit two or more nation states against one another, each equipped with *roughly* equivalent technology, at least in the strategic sense. In the great campaigns of Europe and North America from about 1800 through 1945, it was virtually impossible to disadvantage the enemy through means other than the physical dislocation of his ground forces through maneuver. There were, of course, many small-scale exceptions to this rule, but in general, history's formative lessons for modern Western military thinking emphasized ground maneuver as the precursor to victory.

By removing the principle of maneuver from its limited historical context, we are misapplying it today. We live in an era in which technological change is rapid, and the opportunities for gaining the advantage in combat are many, going beyond the mere movement of units on the battlefield. Further, our warfighting doctrine should address *all* types of conflict, not just warfare against a symmetrical opponent, where classical maneuver ideas apply. This is not to discount the importance and effectiveness of maneuver, but rather to gain a more balanced approach to the overall goal. For these reasons, the principle of maneuver must be revised. The end state of advantage endures, but the means to that end have irrevocably changed and grown beyond the Industrial Age notion of ground maneuver.

It is this failure to distinguish between maneuver-as-an-end-state and maneuver-as-a-means that has distracted and derailed many prominent maneuver theorists today. As historian Daniel Bolger noted in *Maneuver Warfare: An Anthology,* modern proponents of maneuver warfare tend to fixate upon exclusive, favorite examples from history—primarily the German Wehrmacht of World War II. In practice, it is a capital mistake to focus upon one example of something that worked in the past and then build a body of conclusions based on that example. Modern maneuver-warfare advocates often point to the German ideas of decentralized command and control (most often expressed in the German term, *Auftragstaktik*) as being fundamental to their theory. This is an error, because what we are after is *the effect on the enemy,* not the temporarily successful means of achieving that effect. In order to keep clear of the transient and fleeting specifics of what works on the battlefield and what does not, the student of maneuver should stay rigidly focused on the end state desired: a disadvantaged enemy.

What, then, should we do with this principle?

First, we should recognize and emphasize the pursuit of advantage in conflict. I have considered recommending the revised principle be called "Advantage." Consider two insights from past thinkers:

SUN-TZU: With whom lie the advantages derived from Heaven and Earth?

MACHIAVELLI: Wise generals never come to an engagement but when they are either compelled by downright necessity, or can do it with great advantage.

If I were to compose a list of general principles for the instruction of cadets or civilians who were just beginning their study of warfare, I would begin the list with the principle of "Advantage." For the beginning student, we could enumerate the following points about the principle:

1. The many variables of combat will lead, most often, to one side having the advantage over the other. Rarely will two contestants in war be equally matched at all times.

2. The skill of the leader, then, is to first understand the advantages that accrue to him and those that accrue to his opponent.

3. Next, the leader must become adept at *creating* the advantage for the friendly force through the manipulation of the factors of combat. And having created the advantage, he must then *exploit* that advantage skillfully to defeat the enemy.

4. Finally, the leader must learn to offset any advantage that the enemy has.

Such would be the principle of advantage for the novice. But for the purposes of developing a comprehensive revision of the current principle of maneuver, we must go beyond these simplicities into the concept of "Dislocation."

DISLOCATION

An enemy force, in any situation, has strengths and weaknesses. A particular strength of an enemy consists of two elements: a *component* of the enemy's force, and a *condition* in which that component operates. To take a simple example, consider an ancient warrior. He stands before us equipped with a huge sword, a javelin, a breastplate, and a helmet. A fearsome sight indeed! But let us consider this soldier's strength.

The components of his strength are his sword for close combat, and his javelin for the long-range fight. For protection, he has covered his chest and head with armor. But these components are relevant only to a particular kind of fighting. They are designed to be employed in certain restricted conditions—apart from which the weapons and defenses are meaningless.

The javelin, for example, is intended to be thrown. The javelin is an effective weapon when accurately thrown at an enemy soldier about thirty feet or so away. But how does it do against a walled city? Or in a dark alley? Or in a jungle? Is it useful to the warrior who has closed to within knife range with an enemy? What if the warrior is attacked from behind?

If we extend the argument further, we will begin to see the relevance of the javelin fade considerably. How effective is the javelin when our warrior is debating a political point with local government officials, or seeking legitimacy within a political culture? How does it perform for the soldier as he negotiates a business deal with a merchant? Is it useful to him as he tries to win a mate?

All these examples—far removed from the vision of battle that we first conceived—are conditions of conflict that our warrior must face. But in these conditions, the javelin is of severely limited utility. In a similar manner, a given enemy strength always consists of the physical components employed in a very restricted condition. Understanding the *conditional* strength of the enemy is a central part of the art of war, because it leads to the discovery that there are two ways to defeat an enemy's strength: *destroy* the components, or *change* the conditions.

As a modern example, a given enemy force might be especially strong in indirect-fire capability. The primary component of that strength would be the enemy's artillery. The condition of the strength would be the use of that artillery in indirect fire. In other words, the enemy's guns, munitions, sensors, communications, and command facilities would be the physical pieces that compose the strength. But those components operate effectively only when the conditions are right for indirect-fire missions. If permitted to engage

How do you defeat a javelin?
Confrontation or Dislocation?

**Physically
destroy the
components . . .**

**. . . or change
the conditions
of the conflict
to disallow
javelin fighting.**

Figure 3-4

in long-range battles suited to indirect fire, the enemy will be successful. But if we can somehow disallow such fighting, then the components cease to be relevant.

So what do we do when faced with an enemy's strength? Obviously, our foes will attempt to employ their strengths against us to do us harm. How do we defend ourselves from these intentions? Conversely, how do we overcome the enemy's strengths when we are attacking him?

The most obvious and direct approach is to *confront* that strength, with the intention of destroying or degrading it. Against our ancient warrior, we might exchange javelin throws with him, and then close for a sword duel. In the modern example, we would aim at physically destroying the enemy's guns or munitions. The confrontation of enemy strength is a viable and important method of dealing with it. But there are drawbacks to confrontation. It is often dangerous, expensive, and ineffective to take on the enemy where and when he is strong. Therefore, there is another approach to dealing with enemy strength: *dislocation.*

Dislocation is *the art of rendering the enemy's strength irrelevant.* Through dislocation, the friendly force temporarily sets aside the enemy's advantages (in numbers, positioning, technology, etc.) and causes those strengths to be unrelated to the outcome of the conflict. Throughout the history of conflict, armies have used various means—technology, organization, and, very often, maneuver—to dislocate the enemy's strength. Once the enemy's strength is set aside, the friendly force is free to attack through the enemy's weakness to bring about defeat. Dislocation is the theoretical foundation for obtaining the advantage in combat.

We have already implied that by removing our ancient soldier from the very restricted conditions in which his javelin and sword are effective, we can make those strengths irrelevant. In our modern example, we may have the opportunity to dislocate rather than confront the enemy's artillery strength. Instead of attempting to destroy the physical components, we might simply attempt to make that strength irrelevant to the outcome of the fight. For example, if we were able to advance into the battle from a direction that the enemy did not anticipate, he might be unable to apply his artillery fires against us. Or, if we contrived to double or triple our approach speed, we could dislocate his artillery strength, because his tempo of acquiring and firing at targets would not match our velocity. Another alternative might be to march toward the enemy through an area that he does not wish to shoot at, like heavily populated urban areas, or ecologically vulnerable locations. In each of these examples, the enemy's artillery remains strong—but that strength is irrelevant to the outcome, because the conditions have changed.

Dislocation is a useful and enduring idea, because its basis is philosophical, rather than strictly military. Dislocation's continuing utility out-performs contemporary maneuver warfare theory, because the former has a solid foundation in the philosophy of defeating the enemy, while the latter has its roots in transitory twentieth-century tactics. Regardless of the technological context, dislocation theory pertains.

There are at least four types of dislocation. *Positional* dislocation renders enemy strength irrelevant by causing the enemy to be in the wrong place, in the wrong formation, or facing in the wrong direction. Obviously, the old principle of maneuver was aimed at just this effect. *Functional* dislocation sets aside enemy strength by causing it to be dysfunctional, generally through the application of technology or combined arms tactics. *Temporal* dislocation is the art of rendering enemy strength irrelevant through the manipulation of time, and it is the basis for surprise in war. *Moral* dislocation is the offsetting of enemy strength through the defeat of the opponents' will.

I have introduced and developed these ideas in previous works. In *The Art of Maneuver* I introduced the concept and explained the dynamics of positional and functional dislocation, as well as moral dislocation (using the term *disruption*). In *Fighting by Minutes: Time and the Art of War*, I devoted an entire book to temporal dislocation. The goal in this chapter is to show that dislocation theory is the substance behind the old principle of maneuver, and that the traditional rendering of this idea is inadequate. *Maneuver* is a subset of dislocation.

Dislocation is born of weakness; it is the tool of those who cannot successfully confront enemy strength, but who are determined to win anyway. Dislocation is an expression of human resolve not to give up but to find a way to overcome a challenge.

A thorough discussion of dislocation theory exceeds the scope of this book, but the concept is simple. The immediate need is to replace the principle of maneuver with a wider and more germane concept that focuses on the desired end state of a disadvantaged enemy. This is, in essence, what the principle of maneuver has been about since its inception. By revising and renaming the principle at the end of this book, we remove the problem of confusing the means with the end.

BALANCE

If we proceed with the principle of dislocation, we must avoid the mistake of losing our balance. Warfare is a dynamic activity. It exists as a phenomenon that pits conflicting forces against one another. When forces conflict, there is a natural ebb and flow, action and reaction. Because enemies in war consist of thinking, fearing, and determined individuals, they adapt to each other's successes. The problem of adaptation and consequent diminishing effects will be reiterated throughout this book, because it is central to an understanding of warfare.

The enemy in war will not, as a rule, cooperate with our plans. We may well desire to win and win cheaply by dislocating the enemy's strengths and attacking his weakness, but he can be counted upon to oppose us at every turn. In fact, the enemy is trying to dislocate *our* strengths also. As a result, we will find that in the practical realities of combat, both dislocation and confrontation are required. Sometimes we will be able to render enemy strengths irrelevant. Other times, we will have to confront them instead. Hence, a good theory of war must account for a balance between dislocation and confrontation. Both are needed. Both depend upon one another. Both are components of victory.

Therefore, as we come to embrace the principle of dislocation, the intent must not be to presume that all future warfare will feature the dislocation of strength. Rather, it will consist of a combination of confrontation and dislocation. In the past, the terms *fixing* and *maneuver* conveyed the same idea.

"Fixing" the enemy means to immobilize him through the threat of destruction. In other words, we engage the enemy's strength in order to hold him in place. If we are fixing correctly, we give the enemy a choice: either stay put and survive, or move and die. Fixing is an economical way of accounting for, confronting, and controlling enemy strength while our dislocating blow develops.

Sun-tzu gave voice to this idea in his concepts of the "ordinary force" (i.e., the fixing force), and the "extraordinary force" (i.e., the maneuver force). Since that time, many soldiers and writers have repeated the idea of holding an enemy force in place while enveloping or otherwise attacking it.

The challenge for us in the twenty-first century is to lift that idea from its context of ground maneuver, and instead apply it conceptually to the complex warfare of the future. We will continue to fix the enemy, but we will do it through means other than strictly ground attack. We will also dislocate the enemy, but our means of doing so will include much more than a flank attack.

Hence, we are left with a balance between *dislocation* and *con-*

frontation. As often as we can, we will seek to dislocate. But we will enable that idea through the use of confrontation to fix the enemy's strength.

COMBINED-ARMS WARFARE IN THE INFORMATION AGE
The most important expression of dislocation in the realm of small-unit tactics is *combined-arms warfare.* Combined-arms warfare is briefly defined as *the employment of complementary weapon systems to achieve a synergistic effect against the enemy.* The idea is that by orchestrating two or more different weapon systems, the commander can achieve multiplicative effects. For example, while fires from a tank might normally destroy three enemy attackers, and a minefield in the enemy's path might kill one attacker, together they destroy not four enemy systems, but seven.

Combined-arms doctrine appears to be one of the very few immutables in military history. From the beginning of recorded history, leaders in war have sought to find the right mix of archers, spearmen, and cavalry. Later, the lance, harquebus, and pike were the elements of victory. In the early modern period, muskets mixed with artillery and cavalry formed the foundation of combined arms. And in the era of World War II and the Cold War, infantry, tanks, and artillery—later joined by attack aviation—were the key to tactical efficiency. Although the elements of combined-arms warfare have changed over the years, the theory behind the mixture has not.

Combined-arms warfare has at its roots the idea of dislocation—specifically, *functional* dislocation. It is an idea that is easy to understand, but easier still to forget for reasons we shall outline. Hence, it is always an urgent task in any armed force to reinvigorate the teaching and study of combined-arms concepts, especially among the leadership. But this important undertaking is of even greater urgency now, because we are moving into Information Age warfare, and the elements of combined arms have changed once again.

How does combined-arms warfare work? Why does it tend to produce synergistic, multiplicative results in battle? The key to understanding this fundamental of the military art is to comprehend *attack profiles* and *battlefield reactions.*

Every weapon system on the battlefield has an *attack profile:* a way in which that weapon inflicts harm on an enemy target. Rifles fire accurate, low-volume, flat-trajectory bullets at the enemy, while a mortar fires high-angle and delivers explosive shells. Tank main guns shoot high-volume, high-velocity, armor-piercing rounds to distances up to four kilometers, while multiple-launch rocket systems hurl wide area barrages beyond line of sight. Each weapon gets at the enemy in a different way; each has a different attack profile.

What happens when we fire a weapon at an enemy? The first great insight that the student of warfare must grasp is to remember that when a weapon fires, it *kills* some targets and causes *battlefield reactions* in others. One of the reasons that combined-arms warfare doctrine is withering away in the U.S. Army is because we focus too much on kills, rather than reactions. No matter how powerful a weapon may be, each time it discharges, it leaves far more targets untouched than it hits or kills. Hence, a weapon system's most important effect

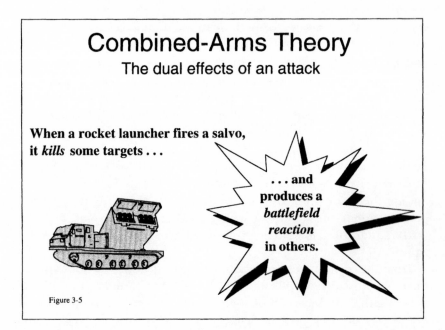

Combined-Arms Theory
The dual effects of an attack

When a rocket launcher fires a salvo, it *kills* some targets . . .

. . . and produces a *battlefield reaction* in others.

Figure 3-5

on the battlefield is not how much it kills, but rather *what reactions it causes.*

When the enemy observes our attacks on the battlefield, he reacts to it. If we shoot a machine gun at the enemy, his reaction is to return fire and seek frontal cover. If he reacts quickly and correctly, the result will be a diminishing of the effects of our machine-gun attack. Hence, when we pulled the trigger, we accomplished two things: We killed or wounded some enemy soldiers, and we caused a greater number to seek frontal cover.

If our tactical capabilities resided completely in one weapon or even one *type* of weapon, we would find that the enemy's reactions were an insoluble problem. After the initial killing we accomplished, our effects rapidly diminish to nothing. We would then have to move some of our weapons to a flank to become effective again.

If, on the other hand, our force has a variety of weapon systems, then we can employ combined-arms dynamics to win the fight. When the enemy takes cover, diminishing the effects of our machine guns, we follow with a mortar attack on his new position. Mortars provide high-angle fire, and the enemy's frontal cover will not protect him from shells lobbed at him from above. Therefore, we can expect to kill some more enemy with our initial rounds of mortar fire. Next, of course, the enemy will react to the mortar attack. The most common reaction to indirect fire is to move out of the impact area. When the enemy moves, we will find, once again, a diminishing of the effect of the mortars (at least until we adjust fire to the enemy's new location). But once the enemy begins to move, he becomes vulnerable to our machine guns again.

This simplistic example is what combined-arms warfare is all about: the complementary effects of weapons. We achieve success by creating combat units with the optimal mix of complementary weapon systems, so that in battle they can capitalize on the reactions caused in the enemy's force.

It is a small step in logic to see how this dynamic applies to other than military endeavors as well. A business might launch a new marketing plan designed to capture market share away from competitors. A sophisticated understanding of interaction, however, leads us to anticipate that a marketing ploy, just like a weapon on the

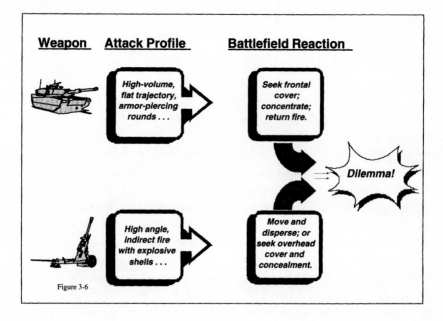

Figure 3-6

battlefield, will have *two* effects. Most immediately, it may result in short-term gains in sales. But it will also cause the enemy (i.e., the competition) to react in a manner that will ultimately diminish our initial success. Provided that we anticipate and plan around these dynamics, we can conceive a sustained strategy of effective competition.

All this is nothing new, but the logic behind it is not well understood or taught in the armed services today. Instead, combined arms has become a political movement as each proponent school or center vies for a bigger share of the budget for its own weapons programs. To get to the right solutions for the future, we must get beyond the budgetary wars and remember the theoretical roots of combined arms. Ultimately, it is an idea aimed at dislocation.

In the twenty-first century, the components of combined-arms warfare will grow and change. The traditional mix of infantry, armor, and artillery, now augmented by attack aviation, will still be part of the equation. But the future battlefield will feature a peculiar addition to the array of weapons: sensors.

Battlefield sensors are growing in numbers and importance. When coupled with precision-fires capability, sensors can give the friendly force the ability to see and understand the enemy's force and engage it at long range. Given that the requisite munitions and firing platforms are available, an enemy force is threatened with destruction if it is acquired by friendly sensors. As in submarine warfare, the actual destruction of the enemy is in some ways anticlimactic. The real battle is about detection.

As we would expect, potential future enemies will seek to avoid detection. Just as with physical attack by our weapons, our sensors also cause reactions—some on the battlefield, others in the laboratory. When the enemy reacts to our sensors, we will find that our ability to detect from any single source will diminish. But, just as in traditional combined-arms warfare, a combination of complementary sensor capabilities presents the enemy with a dilemma.

For example, we decide to employ counterfire radar on the battlefield, and for that purpose, we establish a base line of radar systems just behind our front lines. These radars have a maximum range at which they can detect enemy artillery firing. The enemy begins to fire his artillery, and we quickly respond by detecting the firing units and targeting them with our rocket launchers. After a few repetitions of this, the enemy understands that he must remove his artillery from the range fan of our counterfire radars. As he ceases fire and begins to move farther away, our counterfire radar systems diminish in their effectiveness. However, we also have an airborne radar system that is capable of detecting movement at extreme ranges (like the current Joint Surveillance Target Acquisition Radar System, or J-STARS). When the enemy artillery group was stationary and firing, the airborne system could not see it. But as soon as the artillery began to move, it became visible to the airborne radar. Again, we use this detection to cue fires (or, more likely, to cue other detection assets, such as unmanned aerial vehicles). In this way, a judicious mix of sensor capabilities can force the enemy into contradictory reactions on the battlefield, making him vulnerable to detection and attack from various profiles.

Future enemies will also develop technological reactions to our sensor systems, just as they do against our weapons. Stealth technology is aimed at reducing the detectable signature of weapon systems. The science of stealth technology seeks to first determine the nature of the sensor that is attempting detection. Then, based on the characteristics of the sensor, stealth technology makes the protected system blend into the background. In this manner, stealth tries to make the distinction between the protected system and the environment so small as to defy (or at least delay) detection by a sensor.

For example, one form of stealth technology is aimed at thermal imaging sensors, which detect enemy systems through temperature differentiation. A tank or jeep or aircraft is typically much hotter than the environment it operates in. A thermal sensor can see that difference and convert it into usable data for an observer.

To develop technological protection to thermal imaging, we would have to devise a way to either mask the temperature difference, or actually reduce the surface temperature of the weapon system. These approaches are in fact feasible and used in modern armies. The result, of course, is that the targeted thermal sensors experience a diminishing of their effects.

But just as weapon systems complement each other, so also do sensors, for while counterthermal stealth technology may protect a weapon from detection by a thermal sight, those same measures tend to make that weapon more vulnerable to another type of sensor. The thermal shielding, or the cooling system will create surfaces that are more easily detected by radar, or even by unaided vision. Conversely, measures taken to protect a weapon platform from radar tend to make it more detectable by thermal sights.

One can easily see that the costs to protect a system from all types of sensors would be prohibitive. A device that is undetectable by visual, thermal, radar, image-intensifying, or acoustic sensors would be extremely expensive, to the point that the weapon's primary function of attacking the enemy might suffer. Hence, developers must balance protection from sensors.

On a larger scale, some have suggested that future warfare will feature a violent contest between stealth and data fusion—i.e., the integration of detection data from multiple sources. The idea would be that warfare would pit systems trying to hide from each other, with the winner being the one who can fuse data faster than the enemy in order to detect without being detected. In fact, this type of warfare has already occurred and will continue to characterize future battle. In Operation Desert Storm, the U.S. armed forces demonstrated the effective integration of multiple sensors linked with responsive weapons platforms and key command and control facilities. As the Iraqi ground forces attempted to move and react to U.S. capabilities, they became detectable to a suite of sensors, which led to rapid attack and destruction. Since the early 1990s, these capabilities have improved. The key to prevailing in this type of conflict is to remember and apply combined-arms theory.

COMBINED-ARMS WARFARE VERSUS "SUPER WEAPONS"

Super Weapons plague warfare. They are the enemy of clear thinking and good tactics. What is a Super Weapon? It has two components: a technologically advanced weapon system, and an overly zealous proponent. They are found in every age and, although endlessly discredited in practice, they reemerge from year to year, turning out doctrines and budget wars.

Super-Weapon status is earned by any weapon system that is purported to be impossible to defend against. Some of the most famous Super Weapons of late are explosives and precision-guided munitions. The combination of these two capabilities has created a whole new school of thought: the precision-strike concept—the Super Weapon of tomorrow.

According to its many advocates, precision-strike capability resides in the use of the advanced guided munitions currently in development, along with warheads that can defeat the most resilient armor plating. Several books have been written concerning this latest miracle of military technology, one even proposing that precision strike has ushered in a whole new era. No longer will armies of infantrymen clash on the battlefield. Navies and manned aircraft likewise will be retired. In their place, modern nations will simply shoot very smart missiles at each other. Rather than hammering cities into the dirt with tons of dumb bombs, we will put one Super Weapon through the bathroom window and collapse a society while it's on the potty. Let the revolution in warfare begin!

Precision-guided munitions are a reality; the precision-strike school of thought is total nonsense. The proponents of such thinking have performed a useful service, however: They have shown what can happen to military theory that is unacquainted with military history. Precision munitions are an important part of warfare in the Information Age, but we will ruin their potential if we follow the nonsensical idea that they cannot be defended against. Instead, we must factor them into our combined-arms formula, where we will find them a powerful part of tomorrow's military equation.

The enemy *can* adapt to precision munitions. In fact, a little research and reflection reveals myriad technical and tactical ways to diminish the effects of a guided or smart weapon. Most recently, enemy regimes have deliberately placed noncombatants along with television cameras near potential targets. In this way, they dislocated the possible use of precision munitions by changing the political conditions of the conflict. This technique is just one of many ways to defend against precision weaponry. For those who espouse the precision-strike concept, this vulnerability to countermeasures will spell doom. They will have to cast about for the next Super Weapon. But for those well schooled in combined-arms dynamics, enemy adaptation to precision munitions is no problem. Indeed, we *want* the enemy to adapt to them, because by doing so, he becomes more vulnerable to other methods of attack.

If you believe in Super Weapons, your primary interest is in how lethal your system is, and you must rue the day when the enemy figures out how to defeat your Super Weapon. But if you understand combined-arms warfare, you learn a most startling fact: *A weapon is important only to the degree that its effects can be escaped from.* A paradox of the highest order! Combined-arms warfare does not fear the enemy's reaction to success . . . *it relies upon it.*

This is one of the reasons why nuclear weapons are historically the weakest weapons of war ever invented. It is a matter of record that the destruction wrought by atomic bombs at Hiroshima and Nagasaki fell significantly short of contemporary conventional capability. But even as atomic-weapons technology advanced to modern nuclear devices—even as their explosive yield increased dramatically—their utility to modern war diminished to zero. Quite apart from their utter disconnection from modern political conflict, their biggest disqualification from use in warfare is that they cannot be defended against. They destroy everything, leaving no enemy to react and adapt, leaving nothing to control or possess.

DISLOCATION AND SIMULATION

The doctrine of combined-arms warfare is dying in the U.S. Army, having fallen victim to simulations and doctrinal degeneration. Simulations include computer war games, as well as the live battles that are conducted at the army's combat training centers, such as Fort Irwin, California. These simulations have been responsible for creating the best-trained army in the history of warfare. It is no exaggeration to state that the training simulations first used by the U.S. Army in the mid-1980s have generated one of the most significant revolutions in military capability in our history. Before developing the National Training Center and related training facilities, the army's training was largely unrealistic and not challenging.

Live simulation makes use of laser technology by equipping each weapon system with a laser-emitting device. Each vehicular platform and individual soldier also has laser detectors. In the mock battles that result, weapons fire at each other, and the lasers simulate the actual weapon performance. For example, a tank main gun can hit a target at about three kilometers and kill it, whereas a rifle could reach only about 400 meters and is ineffective against armor.

Army battalions and brigades go through training rotations at these special combat training centers, and they fight against a highly trained, highly motivated opposing force—the "OPFOR." The OPFOR soldiers most often win these battles, because their training level, their knowledge of the terrain, and their experience are unsurpassed. As an infantryman, I have faced real battle as well as the simulated battle against the OPFOR. Real battle was far easier.

The army also makes use of computer war games as training devices. Division and corps commanders and their staffs are trained using these simulations. As with the live simulations, the computer war games use live, highly trained opposing forces. The result is a very competitive, tough training event.

These simulations are a vital part of the army. They have had an inestimable positive impact on America's fighting force, and they should continue to be used and improved. However, there are drawbacks to these simulations.

It is hard to imagine, if you have never experienced it, just how powerful simulations are in the development of today's military leaders. On the surface, every officer and NCO who has participated in simulations admits that the mock battles or computer war games used are just that: simulations. We admit that such devices only approximate aspects of reality; that you cannot use simulations in isolation to understand war; that whatever simulations might teach us must be suspect. We admit these things . . . and then we slavishly obey what the simulations tell us. And while the official policy of these training programs is that no one should overemphasize the simulation results, in practice, the games' outcomes are taken very seriously, even impacting upon career progression in some cases.

Computer simulation pervades all aspects of the armed forces today—not just training. New weapons, as well as new munitions, are developed with great reliance upon computer simulation. No procurement program in the Department of Defense can even be considered without extensive computer modeling to underpin the system engineering and costing.

If computer simulation were used only in the development of future requirements, or in training and leader development, the danger of overreliance would be minimal. But the sad truth is that computer simulation has also influenced our fighting doctrine. Computer war games provide the single greatest influence upon our army's doctrine today.

This assertion would be denied by any official in the army, of course. But while computer gaming does not explicitly feed into our doctrinal development, in fact those games have become gut-level issues to our commanders and staffs. Computer war games are much more than simply major training events. They are among the most powerful formative influences upon our developing leaders. And what is important to the leaders eventually finds its way into doctrine. Some army leaders today have little or no experience in real warfighting, as well as only a superficial interest in military history. By default, their military experience and perceptions are based on *simulated* warfare.

The problem with computer simulation's influence upon doctrine is that current simulations do not approximate real war. Most especially they ignore the moral domain of conflict almost entirely. The OPFOR never routs, surrenders, or flees. They always fight on, regardless of losses. Further, the phenomenon of battlefield adaptation is almost nonexistent. Because computer simulations do not portray diminishing effects in a realistic way, combined-arms theory cannot thrive. This is the reason that combined-arms theory is dying in the armed services. Combined-arms concepts have no reality without the context of diminishing effects and battlefield reaction.

The "never-say-die" behavior of the OPFOR is assumed to be appropriate within the Army, because it represents a tough challenge to the training unit. The most prevalent idea is that the OPFOR must present the "worst-case" scenario against the training force. What could be worse than an enemy who never breaks or retreats? Isn't this the most challenging training experience a unit could face? Absolutely not! The "no-retreat, no-surrender" rule is simply intellectual laziness and completely inappropriate, and this attitude adversely impacts on the army's doctrine and training.

If the OPFOR refuses to retreat, it is, in effect, refusing to train the army in pursuit operations—in practice, the toughest mission of all. Pursuits require fast, agile units that can move resolutely in uncertain situations and reassemble for battle quickly when the next line of resistance is met. The American army has always been weak in pursuit operations, and failures at Manassas, Falaise, and the most recent Gulf War have cost lives, treasure, and political will.

If the OPFOR continues to fight in the face of severe losses and never routs or surrenders, it is teaching American soldiers that to *defeat* the enemy, they must *destroy* them totally. This is completely false. The art of war consists largely of learning how to defeat the enemy *without* complete destruction.

Furthermore, we must consider what the "no retreat" policy does to our logistics concepts. In real battle, enemy units have been known to flee or surrender after as little as 5 or 10 percent losses. Certainly, only the most cohesive units could endure a battle in which they suffered 30 percent losses. Yet in the mock battles and computer war games the army uses, units routinely lose up to 90 percent of their men and equipment—*and keep on fighting!* What does this do to our logistical doctrine? It requires the training force to destroy virtually every weapon system in the enemy force. This volume of destruction requires enormous amounts of ammunition. Typically, units in training simulations expend disproportionate amounts of ammunition but much less fuel than in real war. As a result, logisticians slowly adapt themselves to these games, organizing our resources around the need to deliver huge amounts of ammunition that, in real battle, would most likely not be needed.

In the Gulf War, my battalion carried truckloads of ammunition that we never expended. But we almost ran out of fuel on several occasions. Few realized then or now that the culprit was the simulations that we use in training and in combat developments. A generation of fighting men had learned false lessons about battle and as a result came close to logistical failure.

One of the best ways to reinvigorate army training would be to change the "no-retreat" policy of the OPFOR. Failure to portray moral realities is not "worst-casing it." It's simply shrinking from the hard issues of real battlefields.

Another obvious example of this problem with computer simulation is the degree to which long-range artillery fires prevail in computer simulation. To anyone trained in military history, the notion that indirect fires—whether delivered by artillery, rockets, missiles, or aircraft—can defeat an enemy armed force is absurd. It has never happened in human history and never will. Why? Because the enemy will adapt to any single dynamic on the battlefield. Battlefield fires have always been a vital component to modern combined-arms warfare, but they have never been decisive on their own. In fact, it is embarrassing how often we have had to learn that lesson in war. And yet, due to computer simulation, we have once again fallen for the science fiction of long-range fires.

The principle of dislocation and its offspring—combined-arms warfare—cannot be seriously studied within current computer simulations. Commercial war games do a much better job of simulating real battlefields than government models, because the former depict the moral domain. Therefore, if we desire to rediscover the correct balance between dislocation and confrontation, and if we want to revive combined-arms theory within the services, we must reform our computer simulation and retrain our leadership in military history. This is a vitally important undertaking, because decisions on future weapons and force structure are being made daily, and most are made based on the false assumptions we have learned from computer simulation.

Conclusion

In summary, the principle of maneuver is outdated. It is fundamentally flawed, because it suggests that ground maneuver is the primary way to achieve an advantage over the enemy. In reality, modern armed forces can achieve the advantage through many different means, such as technology, organization, moral and political preparation, and tempo. Rather than clinging to outmoded ideas on maneuver, we should instead focus on the desired outcome: dislocation.

A proper application of the principle of dislocation recognizes the balance between confrontation of enemy strength and dislocation of it.

The four types of dislocation are *positional*, *functional*, *temporal*, and *moral*.

Finally, we should aim at reviving the art of combined-arms warfare within our armed services. To get there, we must improve our simulation of battle and devote ourselves to studying military history and theory. This is necessary to discover and understand the phenomenon of battlefield adaptation and diminishing returns, which is the fundamental concept that underlies combined-arms theory.

4: Offensive

L'audace! L'audace! Toujours l'audace!
—Frederick the Great

Invincibility lies in the defense; the possibility of victory in the attack.
—Sun-tzu

Offensive: Seize, retain, and exploit the initiative.

It is a curious phenomenon that among professional military officers, the offensive aspect of conflict receives disproportionate emphasis over its theoretical counterpart, the defensive. Indeed, within the ranks of the soldiers with whom I have served, this principle has a strong emotional content . . . almost as if an officer who studied defensive tactics too much was of questionable loyalty or manliness!

But a serious study of war leads the student in an altogether different direction. It is pointless to urge upon a general the goodness of offensive operations, when in real war he must always balance offense with defense. As Goltz noted in *Conduct of War:*

> Text-books which discuss the advantages and disadvantages of attack and defence [sic], frequently produce the erroneous impression that a general has a free hand as to which he will select. This, however, will practically never be the case; on the contrary, his action will always be regulated by higher considerations, which dictate the course to be followed by him.

The principle of offensive has never been a valid principle of war. It is an idea that is strewn with problems both in theory and in practice. We will look briefly at the issue of application, and then search out the theoretical flaw of offensive. This inquiry is serious business, because belief in this old principle has led commanders to embark on foolhardy ventures in war—distractions that cost lives and sometimes national sovereignty. In order to come to the truth, we will have to first knock down a series of false assumptions about attacking in warfare.

Probably the most obvious problem with this principle is that of determining what level of conflict it addresses. In my time in the army, I have seen wars, campaigns, battles, and even skirmishes subjected to analysis in the light of "Offensive." Since most of us are not charged with the supreme responsibility of planning and conducting an entire war, it is common practice to apply the principles of war as if they were equally valid as principles of campaigning, principles of operations, principles of tactics, and even principles of small-unit techniques. Consequently, military leaders ranging from General Pétain at Verdun to Second Lieutenant Doe at the National Training Center have been castigated for failing to use the principle of offensive.

How, where, and when are we to apply the principle of offensive? Does this principle applaud or condemn Belisarius, who prosecuted an offensive strategy through a series of tactical defenses? Conversely, how does it pronounce upon Manstein who employed vigorous tactical attacks in order to defend against the attacking Soviets? If offensive is a principle governing all aspects of warfare, are we to conclude that the only acceptable form of military action is an offensive war featuring nothing but offensive campaigns which employ offensive tactics? Surely not!

We know that, in practice, warfare includes both offensive and defensive operations. It would be difficult or impossible to find a war in which both the victor and the vanquished did not employ the defense in their strategy, operations, and tactics. Of course, the principle of offensive, as quoted in the army's capstone manual, FM 100–5 (1993) does allow for the defensive, but only as a temporary measure. Almost as if it were a shameful act!

Commanders adopt the defensive only as a temporary expedient and must seek every opportunity to seize the initiative. An offensive spirit must therefore be inherent in the conduct of all defensive operations.

But if practitioners of war grudgingly admit to the existence of and need for defensive operations, then how can offensive be a principle of war? If it does not cover all aspects of war, then it cannot be considered a comprehensive guide. In truth, warfare, both in theory and in practice, combines the offense with the defense at all levels of war. A distorted emphasis upon the offensive in the French plans for World War I resulted in 300,000 French casualties with nothing to show for them.

The historical context in which the principle of offensive was developed bears on the problem. From 1805 through 1990, the Western world contemplated the phenomenon of total war. The ambitions and frightening determination of Napoleon, Hitler, and Stalin forced their enemies to fight for their very lives. Both in hot war and cold war, Western military strategists had to contend with the idea that two blocs of powerful nation-states were about to enter a life-and-death clash which would end only when one side utterly defeated the other side's military capacity. In this struggle, both sides would have to orient the entire nation's economy, people, and spirit in order to prevail. Likewise, they would have to totally destroy the capacity of the opponent.

In the context of total war, the principle of offensive is a viable strategic consideration. The logic is simple: *There will be no political resolution short of the destruction of the enemy's armed force.* Therefore, it is incumbent upon the general—if he desires to be successful—to relentlessly seek and destroy the enemy in offensive warfare.

But when removed from the specter of total war, the principle of offensive fades in significance. Total war, both in theory and in practice, is a relative anomaly in history. Most armed conflicts come to a political decision long before the actual destruction of enemy armed forces. It is fair to say that the dreams of Napoleon and Hitler expired only after a disproportionate percentage of their soldiers were hunted down and killed or captured. But most other world leaders in history have conceded the point without requiring such extreme measures.

At the strategic level of war, the principle of offensive is anachronistic. The U.S. armed forces will surely find themselves in twenty-first-century conflicts in which they must replace the fascination of offensive with the complex realities of attaining political ends through a judicious application of military means. To instruct future generations of warriors that they must destroy the enemy armed forces through immediate and relentless attack as a prelude to victory is simply in error. Real military operations in the twenty-first century will not comply with such easy formulation.

But even more to the point, we must scrutinize the theory that underlies the principle of offensive: that through offensive operations, one can obtain the initiative.

Offensive operations are the means by which a military force seizes and holds the initiative while maintaining freedom of action and achieving decisive results. This is fundamentally true across all levels of war. [FM 100–5, 1993]

Is there a connection between initiative and offensive action? The word *initiative* comes from the verb "initiate." The noun form of this word is defined as "the action of taking the first step or move." By implication, when we "take the first step," we compel the enemy to abandon (at least temporarily) his own plan of action and instead react to our own. Hence, the idea of initiative envisions two competing wills and advises us that our best chance of winning comes when we preempt the enemy's actions by acting first.

This train of thought is logical and certainly has application to warfare. But is it necessarily related to the offensive? Does offensive action automatically confer the initiative to the attacker, as the principle suggests? Definitely not. Attacks can be slow, uncoordinated, misdirected, and anticipated with horrifying results. The French attacks at the opening of the Franco-Prussian War in 1870 are a clear example of offensive action surrendering the initiative to the enemy.

By the time of the opening battles, both armies were advancing with offensive intent. But due primarily to superior German war planning and logistics, the French rapidly lost all initiative and fell back. In the battle of Weissenburg on 4 August 1870, the French commander, Marshal MacMahon, employed offensive tactics by hurling his cavalry against the attacking Prussians. The result was shocking French casualties and, again, a loss of the initiative. This pattern is repeated in virtually every war in history. Offensive operations fail a lot, and attackers frequently lose the initiative they seek.

During the prolonged Chinese civil war between the Nationalists and the Communists, Mao Tse-tung was most often on the defense, but he retained the initiative almost without interruption. The few instances in which Mao clearly lost his freedom of action came as a result of failed attacks.

Initiating operations and attacking are *not* the same thing, although they may appear to be so. The symptoms of a cold may at first resemble those of malaria, yet the former is a simple malady, while the latter is far more profound and deadly. Likewise, an army that initiates military conflict might appear to be the attacker, while careful diagnosis would reveal otherwise.

The connection between initiative and offensive operations is spurious and hard to discern. Without a clear theoretical understanding of the difference between offense and defense, we can succumb to the utterly false idea that whoever moves first is the attacker. In reality, the roles of attacker and defender are related solely to the *objective* of a given operation. The defender possesses the objective; the attacker strives to obtain it. If this assertion is true, then we can see that the matter of "initiative" (i.e., who moves first) is unrelated to offensive action.

For this reason, I believe that the principle of offensive is theoretically flawed: There is no causal connection between initiative and offensive action. The supposed relationship is illusory and leads us to believe in (or at least proclaim belief in) one role (offense) as being most often preferable over the other (defense). The so-called principle of offensive has led past commanders to commit the grossest errors, costing thousands of lives. Further, Clausewitz pointed out that the defense is the stronger form of combat, and his assertion is still true, because attacking requires a combination of movement and striking—activities that trade off against each other. This is not to say that we should reverse course and embrace the defense as a better form of combat. Rather, the lesson is that practical experience in warfare teaches a balance of offense and defense.

JOSHUA CHAMBERLAIN MEETS SADDAM HUSSEIN

I am ashamed to admit that as a child, I enjoyed telephone pranks. There was nothing more exciting than calling an unknown person and pranking them by asking them if their refrigerator was running ("Better go catch it, then!"). Telephones were a great source of entertainment when I was a kid.

As an adult, I have realized how misbehaved and ill-bred telephone pranksters are these days. It was cute when I was a kid; now I think it should rate some serious jail time for the offenders. Up until a few years ago, kids pulling telephone pranks were a real nuisance.

Enter Caller ID. When I bought this new capability to identify callers, I literally waited in ambush for a prank call. It took a while but it happened. Some nasty little kid asking me if I had Prince Albert in a can or something. But I could almost hear his heart valves lock up when I slowly repeated his last name and phone number to him. Life in the Information Age is a hoot!

Telephone pranking in the context of today's technology simply isn't feasible anymore. The risk is much too great, because the target is aware. The same dynamic applies on the modern battlefield, except that the prank we used to pull was given the honorable name of "initiative."

On 2 July 1863, the Union left flank at Gettysburg was in grave peril. Following unsuccessful attacks and indecision the day before, the Army of Northern Virginia was about to attack the Federal lines again. This time, however, General Lee envisioned an envelopment against the undefended—or, at any rate, weakly held—Union left flank. After interminable delays, Lee prevailed upon Longstreet to send the divisions of Hood and McLaws against the Union left.

Fortunately for the Federal cause, General Meade's engineer, Brig. Gen. Gouverneur K. Warren, spotted the potential weakness

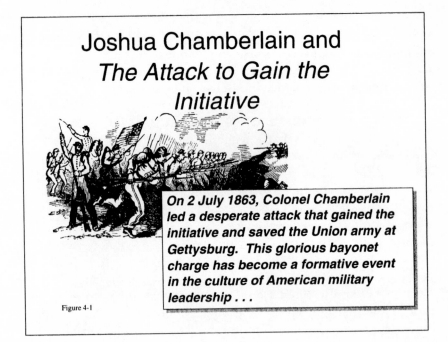

Joshua Chamberlain and
The Attack to Gain the Initiative

On 2 July 1863, Colonel Chamberlain led a desperate attack that gained the initiative and saved the Union army at Gettysburg. This glorious bayonet charge has become a formative event in the culture of American military leadership . . .

Figure 4-1

in the Union lines and directed some nearby troops to occupy the key high ground at Little Round Top. In command of the extreme left flank of that position was a professor-turned-soldier named Joshua Chamberlain.

Chamberlain was a brilliant, energetic, and resourceful man, and on that fateful day, he well understood the importance of the position he held. If the Confederates were able to turn him out of his position, the entire Union line might collapse. Chamberlain determined that his Maine regiment would not fail.

As Gen. John B. Hood's soldiers advanced against his position, Chamberlain directed a spirited defense. But in the last crucial moments, when his soldiers were almost out of ammunition and Hood's division was preparing its last desperate assault, Chamberlain made the most important decision of his military career.

He ordered his regiment to fix bayonets and charge.

As Hood's division climbed toward the Union lines for one final attack to seize Little Round Top, Chamberlain's men fell upon them in a countercharge that swept the Confederates back down the hill. The Union left held. Joshua Chamberlain had saved the day.

Within the ranks of the U.S. armed forces, this dramatic incident is among the best-known anecdotes from American military history. In fact, Chamberlain's inspired conduct that day is cited in official doctrine on leadership. It is also commonly referred to as proof of the principle of offensive. And rightly so. Is this not a clear example of a well-led attack seizing the initiative away from the enemy?

Indeed it is. In fact, it is a perfect example, because it goes far in showing why this "attack-to-seize-the-initiative" idea is losing relevance in modern operations.

What do we learn from Joshua Chamberlain's magnificent charge at Little Round Top? If we are interested only in a superficial understanding, then we may comfortably repose in the classic explanation that the attack leads to gaining the initiative. But if we are interested in the real truth of the matter, then we must understand the essential context of that fateful charge. For in reality, Joshua Chamberlain had with him a secret weapon that day.

Chamberlain's secret weapon was *bluff*. It was not the tangible, active kind of deception common in military planning today. Instead, Chamberlain's entire concept for the charge was founded upon his deception—both of the enemy, and of his own men—perhaps even of himself. The reason Chamberlain's charge worked that day was because both the Union and Confederate soldiers interpreted the maneuver as a demonstration that Chamberlain's unit was stronger than their Confederate opponents. In the excitement and confusion of the clash, the soldiers in gray believed their lives to be in peril before the onslaught of a superior fighting force. The sudden charge disguised the true condition of Chamberlain's regiment. Little Round Top was saved by ignorance, which led rapidly to moral collapse.

Consider how little Hood, McLaws, Longstreet, and Lee knew about the Union left flank that day. Of these four men, John B. Hood had the most accurate read: He knew that Little Round Top was defended, because his scouts had reported that fact to him earlier. Longstreet almost certainly believed the position to be held, but he allowed himself to be overruled by Lee's insistence on an attack. At best, the Confederate attackers that day had a vague impression of what lay atop the summit as they climbed.

Likewise, Chamberlain knew little of the situation, except that the Confederates were determined to win the position—an eventuality that he could not allow. Having received several charges earlier, Chamberlain could not have known with any certainty that Hood's division was close to culminating.

And so into this admixture of fear, rumor, suspicion, and above all, ignorance, Joshua Chamberlain led a bold attack that secured the Union flank and his own place in history. What must we conclude from this feat?

Initiative is 90 percent bluff—and as such it works only in the context of ignorance. The attack to gain the initiative, when it is vectored against a blind, misinformed enemy, is a viable technique and, on occasion, a battle winner. But what happens when the enemy is aware? What happens when a Joshua Chamberlain charges into an armed force that knows the true situation?

On 22 January 1991, a battalion-size convoy of Iraqi armored fighting vehicles was heading south toward a small town in Saudi Arabia named Khafji. Within minutes, an E-8A Joint Surveillance and Target Acquisition Radar System (J-STARS) aircraft detected the movement and guided Coalition combat aircraft to intercept. Soon after, fifty-eight burning wrecks lay scattered on the desert floor, with the shaken survivors racing back to safety.

One week later, Saddam Hussein was ready to try again. Frustrated at his army's inability to act under the weight of the U.S.-led air campaign in the early days of Operation Desert Storm, Hussein conceived a plan to seize the initiative from the Coalition through an attack into Saudi Arabia.

In many ways, the Iraqis' abortive attempt to wrest the initiative away from his enemy is reminiscent of Chamberlain's dilemma at Little Round Top. Hussein realized that to allow the enemy to fight according to their own timetable would eventually result in his defeat. Like Chamberlain, Hussein recognized the value of a well-timed attack. And, like Chamberlain, he had only a vague understanding of the numbers and disposition of the enemy.

But that was where the similarity ended. Hussein, like many leaders before him, was going to try to bluff his way into the driver's seat with his attack on Khafji. Unfortunately for Saddam Hussein, warfare had dramatically changed, and with it, the nature of initiative.

On the evening of 29 January 1991, Hussein's 5th Mechanized Division, with support from sister units, launched a poorly coordinated attack across the border between Kuwait and Saudi Arabia. Two of the three attacking brigades were turned back quickly by Coalition forces, but the third managed to reach Khafji, which it occupied with a mere battalion. A counterattack by an Arab force eliminated the isolated Iraqis several days later, and Hussein's "lightning strike into the kingdom of evil" was over.

Beyond the obvious loss of men and materiel, the attack served only to impress upon Coalition leaders how inept Hussein's army really was. Although impressed somewhat with the spirit of the attackers, U.S. general Norman Schwarzkopf and his staff immediately recognized Khafji for what it was: a clear demonstration that the U.S.-led Coalition had the initiative and would retain it.

If offensive operations are the best way to seize and maintain the initiative, then why did it accomplish just the opposite for Saddam Hussein? The reason is that the Iraqis attacked *without the context of ignorance*. The Coalition forces had a comprehensive view of the battlefield. Although their knowledge of the Iraqi dispositions was by no means perfect, it was good enough not to be bluffed so easily by what amounted to a disjointed divisional reconnaissance-in-force.

So if we want to be unthinkingly conservative and embrace this aged principle of offensive, what are we to conclude? That it leads to the initiative only *sometimes*? Perhaps it works for some nations, but not for others? Maybe it only works on certain days of the week, or when the moon is full . . .

Or maybe there is a fundamental flaw in the logic! Read on!

INITIATIVE IN THE INFORMATION AGE
In the past, initiative has been analogous to chutzpah: It seemed like a good thing to have, but how did you get it? Because we did not properly identify the true but disreputable nature of initiative as being essentially a bluff, we instead elevated it to a mystical intangible,

and then sought to identify reliable methods to attain it. To wit, we decided that offensive operations produce initiative, and we crafted this proposition into a lofty principle of war.

It's all hogwash.

Offensive operations do not give the initiative to anyone. The relationship is entirely illusory. Initiative is the offspring of bluff, and when removed from the context of battlefield fog, it's like a fish out of water. The Information Age most certainly will continue to feature such battlefields—i.e., battles in which one or both contestants are fundamentally ignorant of what is going on. In such contests, the attack to seize the initiative will continue with its dubious but much-vaunted tradition of success.

But the Information Age will also produce a new phenomenon: the aware enemy. Hopefully, it will continue to be *our* side that has the truth. But in either case, we must understand that on battlefields and in wars in which one side or both can see and understand the situation with accuracy, initiative must be redefined.

THE "I" WORD

Language is supposed to facilitate communication, but as anyone who has teenagers has learned, words and thought are often disconnected. I have learned from my sixteen-year-old son that "bad" means "good," and "fat" means . . . well, something other than what I thought it meant. Some words actually *inhibit* communication, rather than advancing it. Such is the case with the principle of offensive. The central reason that this principle has misled us in the past is because of its connection to the term *initiative.*

Initiative is a bad word, and we should drop it from military parlance. It is a word that means one thing, but appears to mean something else. As previously mentioned, the word is the noun form of "initiate." The dictionary definition of the word is all about *starting.* Therefore, by starting something—anything—you ipso facto have initiative.

But is this what we really mean when we talk about having the initiative? When we envision an army sweeping forward through a theater of operations, with the enemy in headlong retreat, we say the victors have the initiative. It's not just a matter of starting to move or attack. Rather, we are pointing to a sustained freedom of action—a continuous opportunity to move and attack.

Is initiative something you *have* or something you *do?* The problem with the word is that it's really about *doing* something, but we use it as if it's a commodity we can own. Because of this confusion, we have learned the false lesson that by *initiating* the action (which in military terms sounds a lot like attacking), we will gain and maintain this thing called *initiative.* Confusing.

For this reason, we should avoid the "I" word completely. What we really want in warfare is freedom to act, or, to put it another way—*opportunity.* Opportunity is a much stronger concept than initiative, because it connotes potential freedom of action *throughout* an operation, rather than just at the commencement. A commander who seeks initiative has only one recourse: move or attack now! This is not a wise thing to do. Premature movements or attacks lead to disaster. On the other hand, a commander who seeks opportunity has many options to achieve it and sustain it—including attacking—but also many other options.

Opportunity in war comes about by increasing friendly capability and decreasing enemy capability. A commander who builds forward logistics bases increases his opportunities for action. An army that has high morale has greater opportunity than one whose spirit is sagging. A corps with robust engineer capability has greater opportunity for movement and attack than one without. A force that has achieved a high degree of political penetration of a theater of war has greater opportunity than one that is alienated from the population.

Opportunity is a broad concept of interaction with the enemy. The old concepts of initiative that embodied the principle of offensive were narrow and relevant in few scenarios. Following the old adage of attacking to gain the initiative will most often lead to failure, not success. If future commanders instead shift their focus to the real issue—opportunity—they will be free to develop more varied approaches toward sustained freedom of action.

Attacking the enemy to create relative freedom of action (i.e., the old principle of offensive) can still work and should still be taught—but only as one of many ways to obtain opportunity. A good chess player knows that you set up a winning game not only by physically attacking the enemy to gain transitory tempo advantages, but that there are other dimensions to creating opportunity, as well. Positioning rooks, castling, developing minor pieces, and protecting pawn structure are all aimed at widening the scope for decisive action in the future. They are as much a part of attacking the enemy as the physical exchange of pieces. In fact, it is these strategic moves that distinguish an expert player from a beginner, who usually is focused on immediate indecisive attacks and exchanges.

The other side of opportunity is reaction. Because war is the interaction of opposites, we know that the enemy will be vying for opportunity just as we are. When he has it and exploits it, we will have to react to his actions. The goal must always be to react in such a way as to recapture the opportunity for the friendly force. A tired and demoralized commander who is largely ignorant about the battlefield might react just enough to survive. This type of reaction is not sufficient for victory in war. Instead, commanders must view every enemy action as a chance to seize freedom of action from the enemy.

In this sense, warfare is truly a contest of wills. When the German army attacked in the Ardennes Offensive in 1944, most Allied officers viewed the attack as a disaster to be defended against. Their vision went no further than surviving the onslaught. But a select few viewed the developing salient as an opportunity for inflicting a decisive defeat upon Hitler's war-making potential once and for all. In the deepest, darkest moments of fear during war, the great commander sees opportunity and acts to bring it about.

Conclusion

The principle of offensive is inaccurate and not fit for duty in the twenty-first century. The logic that underlies it is suspect, and its history of success dubious at best. Commanders who heed it, beware! Many of your forebears brought about personal and national disaster with their ill-considered attacks.

There are better ways to think about interacting with the enemy than crying *"Toujours l'audace!"*

We must have the courage to look at the aged and respected commodity of initiative and realize its true nature. It is, as we have seen, largely about bluff, and it works and has relevance only in the context of ignorance. When used against an aware enemy, it fails with almost statistical certainty. Future warfare will reveal—either through intellectual reasoning or bloody failure—that initiative is a weak, limited concept. It is better replaced with concepts of Opportunity and Reaction.

In the end, we will discover that, as with dislocation and confrontation, *balance* is the key. Future warriors must learn the power of both opportunity and reaction, and learn to combine the two for dynamic effect.

5: Mass

Multitudes serve only to perplex and embarrass.

—Saxe

There is no higher and simpler law of strategy than that of keeping one's forces concentrated. No force should ever be detached from the main body unless the need is definite and urgent.

—Clausewitz

Lord, it is nothing with thee to help, whether with many, or with them that have no power: help us, O Lord our God; for we rest on thee, and in thy name we go against this multitude.

—King Asa at Mareshah, 2 Chronicles 14:11

Mass: Mass the effects of overwhelming combat power at the decisive place and time.

Mass is not a valid principle of war in the Information Age. In fact, it has never truly been a valid principle, as we shall see. But it has been my practical experience that unfounded devotion to "mass warfare" is a positive roadblock in developing modern warfighting concepts. Of all the principles of war, the American soldier has the most trouble with this one: clinging to a colossal anachronism while the future of warfare beckons him to drop it and move forward.

In order to see why this principle is so ill suited to guide us into the future, we will first look at why mass has never been valid as a principle of war. Next, we will examine the classical arguments in favor of mass and see how those arguments no longer apply. Finally, we will see how mass trades off against precision, velocity, and acceleration in war.

One note of caution: Readers familiar with current military doctrine will wonder why I am not using the phrase "massing effects," as our field manuals do. Rest assured that we will discuss the idea of massing effects at the end of the chapter. But before we take on this latest modification of the principle, we must first deal with it as it evolved historically.

HISTORICAL INVALIDITY OF MASS

To begin with, we must see why mass has lacked validity in the past. Someone once noted that "We write to refute." In other words, whenever someone publishes a thesis declaring something to be true, he is, in effect, *refuting* a previous, perhaps tacitly understood concept. For example, if I wrote a paper espousing the use of synthetic lubricants in engines, the point would be to argue against the use of petroleum-based products. I would be refuting a common practice. If this were not so—if everyone already agreed with the thesis—then writing it down would be superfluous, and it probably would not be published.

Likewise, in our search for the truth about mass, we must begin with the epistemology of the idea. Military thinkers long ago began to frame the foundations of the modern principle of mass, because they were refuting a contrary practice: that of dispersing one's army. There would be no point in reiterating to future generals the importance of keeping one's forces concentrated unless those generals would be faced with tempting opportunities to forego such concentration. And, indeed, there are and always have been *many* reasons not to mass.

Armies on the move tend to disperse and assume movement formations, rather than fighting formations, typically in columns along roads. If the problem of movement is large enough, the army will disperse along numerous road networks for efficiency. Armies tend to "unmass" during movement. Moltke the Elder noted that armies massed for battle have temporarily lost all capability for operational movement. Conversely, those armies conducting large-scale movements are ill disposed to fight.

Movement is a natural and valid function in war. If, in an attempt to stay true to the principle of mass, we disallowed any movement,

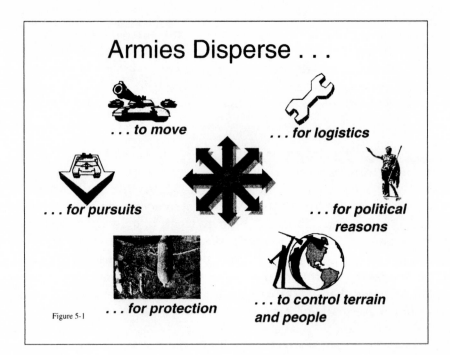

Armies Disperse . . .

. . . to move

. . . for logistics

. . . for pursuits

. . . for political reasons

. . . for protection

. . . to control terrain and people

Figure 5-1

there would be no warfare. One can imagine two opposing generals, both trying to stay massed, glaring at each other and unable to close the distance for lack of movement capability.

Armies also disperse for logistical purposes. Prior to the Industrial Revolution, armies frequently had to forage. Because foraged supplies were limited, the entire army or corps could not subsist in the same general location but had to spread out to live. Likewise today, logistical support takes room: A refitting army or battalion is not able to fight. Although it may not be massed for battle, it is conducting perfectly legitimate wartime operations.

As another example, armies disperse to control terrain and population. People do not live on battlefields, but in towns, cities, and farms. As the objective in warfare frequently includes enforcing a policy within a population, armies have often found themselves dispersed in that function. A massed army cannot compel obedience, monitor treaty obligations, or control terrain.

With the advent of the rifle, rapid-firing artillery, the machine gun, and modern weapons of mass destruction, armies have gradually dispersed for protection. Concurrently, as soldiers replaced spears with muskets and, later, semiautomatic weapons, there was no longer a need to mass. Instead, dispersed soldiers could concentrate their *fires* rather than the points of their spears. Hence, at the tactical level of war, there was both an opportunity to disperse and an urgent obligation to do so for force protection.

Although American thinkers have perhaps forgotten it (given our cultural assumptions), armies of the past have sometimes dispersed in order to avoid armed rebellion. It was a fundamental part of imperial Roman strategy to refrain from massing too many legions, especially when a charismatic or influential leader was in charge. The assembly of too many legions could (and often did) lead to a rebellion among the troops—who might then march on the capital or even establish an independent state. Part of the tension in the Roman opposition to the barbarian incursions resulted from a need on the one hand to mass for battle, and the need on the other hand to disperse for political reliability.

Finally, armies disperse to conduct pursuits and exploitations— primarily because such operations require rapid and economical movement, and because the enemy is temporarily incapable of opposition. The general who insists upon remaining massed will be one who utterly fails in pursuit operations, and the U.S. Army has frequently demonstrated this weakness.

With so many occasions to disperse or at any rate to dispense with concentrations, how can we claim that such concentrations compose a principle of war without at least allowing for the opposite idea? As Goltz noted: "The modern art of war, therefore, involves alternately separating and uniting bodies of troops."

WHY MASS?

If there are so many valid reasons not to mass our forces in war, then what are the justifications in favor of massing? Why have past thinkers perceived a need to concentrate for battle?

In order to understand why mass is no longer valid as a modern principle of war, we will examine each of the reasons that armies have sought to mass in the past.

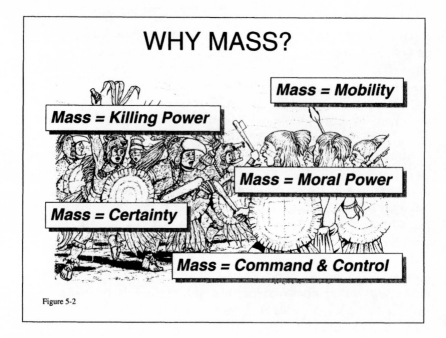

WHY MASS?

Mass = Mobility

Mass = Killing Power

Mass = Moral Power

Mass = Certainty

Mass = Command & Control

Figure 5-2

Mass Equals Killing Power

The most compelling idea behind the principle of mass is that mass equals killing power. That is, by concentrating combat power, the friendly force can kill maximum numbers of enemy weapon systems and soldiers. Concentrating to kill is a natural and intuitive idea—first conceived, we might suppose, as primitive men surrounded the wild boar—and we can point to endless examples in military history to show that this idea has merit. But to advance our understanding of warfare, we must come to grasp *why* mass has conferred killing power.

The reason mass has helped us to kill is that for most of human history, *one man has been able to kill fewer than one of his foes in battle.* In other words, when two armies clashed, each soldier, on the average, killed fewer than one enemy soldier. The instruments of killing—whether bare hands, a spear, a knife, a sword, or a primitive firearm—tended to be so inaccurate, and have so low a rate of destructive blows or shots, that a single enemy death required many repeated attacks—

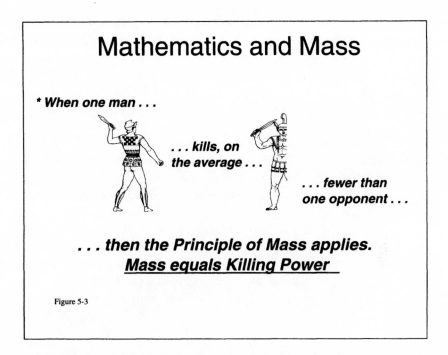

Mathematics and Mass

* When one man . . .

. . . kills, on the average . . .

. . . fewer than one opponent . . .

. . . then the Principle of Mass applies.
Mass equals Killing Power

Figure 5-3

or simultaneous attacks by many soldiers. Those who in practice killed more were heroes—exceptions to the rule—and the subject of great praise. "Saul has slain his thousands, and David his tens of thousands." But the average soldier had to get help to slay even one enemy. The mathematics of battle in the Agrarian and Industrial Ages demanded a concentration of soldiers in order to overcome the incapacity of each individual. The only way to reliably kill an enemy soldier was to have two or more friendly soldiers attack him. Hence, the principle of mass.

But the principle of mass began to lose relevance as military technology progressed. As early as 1861, in the American Civil War, we have clear evidence that mass was unable to keep pace with modern war.

In the 1700s, gunsmiths were developing rifled barrels, thus increasing the accuracy and range of firearms. When, in 1810, the American John H. Hall invented a breech-loading flintlock rifle, the frequency (i.e., rate of fire) of small arms began to rise dramatically.

The famous Minié ball was invented in 1849, and the primitive inaccuracies of the musket ball were replaced with a startling new capability to hit a target with long-range rifle fire, although doctrines based on short-ranged, inaccurate firearms persisted . . . with terrible consequences. Over the course of a single generation, small arms transformed from slow, inaccurate, unreliable weapons into high-volume, long-range, accurate killing machines. The solitary shot of a musket—of questionable lethality even within one hundred yards—became the life-ending crack of a distant, smokeless, invisible rifleman, who could easily destroy ten of his adversaries if they were foolish enough to expose themselves to fire. A concentrated group of charging soldiers ceased to be a means of killing, and instead degenerated into a handsome target.

All of these elements (except for smokeless powder) were in place at the Battle of Manassas, 21 July 1861. And in the ensuing years, both in America and throughout the civilized world, mass infantry tactics began to whither away under a hail of bullets. It took many years to

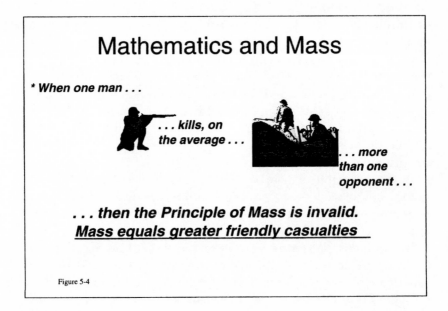

Mathematics and Mass

* **When one man . . .**

. . . kills, on the average . . .

. . . more than one opponent . . .

. . . then the Principle of Mass is invalid.
Mass equals greater friendly casualties

Figure 5-4

overcome the traditional massed charge that had been the ultimate expression of combat power in the days of the pike or bayonet. But eventually, after hundreds of thousands of casualties, soldiers began to let go of tactical massing.

Mass and Armored Warfare

But if small-arms technology began to chip away at the principle of mass more than a hundred years ago, then why did this aphorism survive and even thrive? Because in 1860, one year before the First Battle of Manassas, a Belgian-French inventor named Etienne Lenoir designed the first internal-combustion engine. While infantry-based armies conducted their great struggles for the remainder of the nineteenth century through the beginning of the twentieth century, the internal combustion engine slowly gave rise to the automobile, and, incidentally, to a new engine of war. The tank combined first two, and later three elements that temporarily invigorated the principle of mass, despite its fatal diagnosis in 1861. To begin with, the armored combat vehicle combined protection with mobility. Suddenly, the trenches of Petersburg and Sevastapol were given legs by the internal combustion engine, and they strode across the battlefield of Cambrai (November 1917), collapsing the German defenses in short order.

In the period between the two world wars, tank technology began to add the third component—firepower. By the end of World War II, the principle of mass was in full swing once again. Tanks—each capable of killing, on the average, fewer than one of its armored adversaries—joined in mass tank armies to decide the fate of the world. Soldiers and thinkers relearned the lessons of mass.

But as we contemplate the battlefield of the twenty-first century, we must reckon with a new phenomenon: *that one shot equals one or more kills.* That is, the technology of direct and indirect weapons today allows a single weapon system—under the right conditions—to kill multiple targets with one shot, or to kill with multiple shots in rapid succession. In the coming years, we can expect that the three central components of modern firearms will continue to advance: precision, volume of fire, and lethality. In fact, they have already overtaken (perhaps temporarily) the protective capability of modern weapon systems. The result is that tomorrow's battlefield will feature an unheard-of destructive capability.

But if we can intellectually grasp the "one weapon, multiple kills" potential of future battle, then surely it is a small step to see that the principle of mass has lost one of its strongest pillars: *Mass no longer equals killing power.*

The dynamics of this conclusion are twofold. On the friendly side, there is no *need* for mass, because a smaller group of weapons can kill many targets. On the enemy side, mass becomes only a liability, providing the enemy with convenient and lucrative targets.

In a recent study that I conducted, I found solid statistical proof that in most battles in history, the *smaller* side won. In fact, in 56.5 percent of the battles I collected data on, the side with the fewer soldiers won, while only 36.4 percent of the battles were won by the larger side. The other 7 percent were fought by roughly equal numbers. Now this statistic alone does not disprove the principle of mass. It may in fact validate it. Often, the side with inferior numbers wins through an efficient concentration of force against a portion of the enemy. But the overall trend does make clear that a concentration of numbers does not lead to victory.

Given these changes in warfare, we must at the very least conclude that the nature of the principle of mass has changed. Mass will not convert to killing power in the future. But what about the other arguments for mass?

Mass Equals Shock

Closely related to the "mass equals killing power" idea is a general belief in the relationship between mass and shock. Shock is an ill-defined but very present phenomenon on the battlefield. No veteran of battle can mistake its influence. It is in the eyes of a dazed enemy surrendering; it is the cacophonous music of panic on a radio net; it is a gnawing fear that things are falling apart. But despite the fact that shock has defeated more armies and won more battles than any general could ever claim, it remains an amorphous concept. To investigate the relationship between mass and shock, we must begin by defining this mysterious killer.

First, shock is an emotional state of mind. There are certainly physical manifestations of shock when it occurs: dead and wounded soldiers, destroyed equipment, dispersion, retreat, and so on. But death and destruction do not always bring about shock. Shock occurs in the mind, not on the battlefield.

Second, shock is an emotional state brought about by the interaction of adversaries on the battlefield. The attacker cannot cause shock if there is no defender; the defender won't experience shock unless he's attacked.

Shock is a feeling of numbness, hopelessness, and dismay. It is fear generated from the sudden realization that the plan is falling apart, and that there is mortal danger nearby. When shock occurs, individual priorities change. Loyalty to a plan, a unit, or a leader are replaced with panic or malaise.

The vital component of shock is violence that occurs with such ferocity and tempo that the victim is thrown from a calm, coherent state of mind into a lesser state. It is not so much a matter of how *much* destruction occurs, but rather how *rapidly* it occurs. Shock is worsened by ignorance of surroundings, lack of training and experience, and disconnection from leaders.

If we have the definition of shock about right, it is no large step to understanding that just as mass is no longer a requirement for killing power, the same is true for shock. Modern technology, when coupled with a dynamic and aggressive fighting doctrine, can inflict high-tempo violence upon the enemy without overwhelming numbers. Further, the offensive use of information warfare can exacerbate the isolation of enemy soldiers from their leaders and can worsen the ignorance of the adversary by blinding him.

In short, if mass no longer equals killing power, it ipso facto no longer equals shock either. But are there other reasons to mass?

Mass Equals Certainty

General Colmar Freiherr von der Goltz believed in mass. He was a notable soldier in the Prussian army in the late nineteenth century, and his writing influenced many in the generations prior to World War I. Concerning the need for concentration on the field of battle, he wrote:

> The surest way to conquer the enemy's main army is to concentrate a numerically superior fighting mass, for no one can

assume beforehand that he will have the better general at the
head of his army, or that the latter will be braver than that of
the enemy. Such factors for calculating on victory may certainly
be taken into account when the tasks set us in the field are being practically carried out, and there they will often carry more
weight than any others in enabling us to arrive at a decision.
In the absence, however, of any marked superiority in one side
over the other expert opinion will always suppose troops of
equal quality on both sides. And then, next to the suitability of
the measures adopted by the leaders, numbers are more decisive than anything else.

Goltz wrote with great clarity of thought. On the one hand, he un-
abashedly proclaims his belief in the concept of concentration of superior numbers. On the other, he takes care to explain *why* he relies
on mass, and admits that in actual battle, other determinants are
likely to weigh heavier than numbers.

We cannot ask for any greater circumspection from someone writing at the turn of the century. Mass warfare was a foundational concept in Prussian military strategy at that time. It is remarkable that
a trained officer could find it within himself to question the assumptions behind the principle of mass as applied to Prussian con-
ceptions of war.

Nevertheless, we must question those assumptions. Goltz allowed
for a belief in mass, *because of the uncertainty of battlefield conditions.* By
his own admission, superior numbers served mainly as a hedge
against the unknown. In battle, it was probable, in the author's view,
that other factors—leadership, tactics, morale—would decide the
day. But in the late nineteenth century, those factors could not be
reliably predicted and thus could not weigh in on questions of strategy. Instead, soldiers of Goltz's day compensated for uncertainty with
mass.

Much of the reputation of this vaunted principle lies in its ability
to overcome doubt, ignorance, and uncertainty. Linear tactics were
developed to maximize killing power in the days of spears and rifles.
But the concept of linear formations quickly outgrew the battlefield
in the early modern period, and generals began to spread their

armies across entire theaters of war. In this, they were compelled to dispose their troops to ensure that enemy armies would not surprise them and turn their flanks, or threaten some vital resource in their rear. In order to compensate for the general's inability to know where and when the enemy was coming, he had to pack huge numbers of men on the avenues along which the enemy *might* advance. Mass compensated for uncertainty.

Recent experiments in the U.S. Army have shown that the intuitive use of numbers to overcome uncertainty dies hard. Even when modern commanders have a virtually perfect picture of the battlefield, their instincts incline toward mass warfare. We have been trained and conditioned to shove large numbers of soldiers and weapons at battlefield fog, in order to roll it back or at least mitigate it. If in the future we field technology that can peer through that fog and allows us to perceive the battlefield with precision, then we must train ourselves to abandon the wastefulness of mass. This will not be an easy task.

Perhaps by observing trends in the technical aspects of war, we can induce lessons at the tactical and higher levels. We know now that precision munitions, such as smart bombs, cruise missiles, and guided weapons, reduce the need for massed attacks. In place of armadas of B-17 bombers dropping thousands of dumb bombs, we now use a handful of precision weapons. We have reduced the inaccuracy—the uncertainty—of our long-range strike capability, and as a result, we no longer require large numbers of munitions. In the past, we compensated for the uncertainty of inaccurate weapons with numbers. In the future, as precision replaces inaccuracy, we will reduce the numbers of weapons required.

If this trend is true at the technical level of war, surely it is a small step in logic to see how it applies equally to tactics, operational art, and strategy. When the relative ignorance which pervaded warfare—especially since the nineteenth century—begins to give place to knowledge, information, and truth, the commander should and must replace *mass warfare* with *precision warfare:* the accurate allocation of combat power to achieve a specific purpose.

Naturally, the degree to which we replace mass with precision will be dependent upon how effective information technology is in real war. It is intuitively obvious—although cynics relish declaring it—that the technology we are developing will not always work or reach the high standards we strive for. And, as we will reiterate throughout this work, the enemy will actively resist all our successes, including our ability to see and understand the battlefield through information technology. Therefore, future warfare will fall short of the concept and, consequently, mass warfare will not wither quite as quickly as the theory would have us believe. Precision will replace mass in proportion to our ability to minimize uncertainty.

Mass Compensates for Lack of Mobility

Goltz noted that one of the reasons a contemporary commander needed greater numbers than the enemy in battle was so that he could "overlap" the enemy's flanks. This idea envisions two roughly parallel lines of troops facing each other for battle. Naturally, there is an advantage to the side with the greater numbers (hence, the longer line), because they could then wrap around and envelop one or both of the enemy's flanks.

Even in the comparatively linear tactics of Goltz's day, commanders intuitively understood the need to turn the enemy's flank or otherwise gain the advantage, as we have discussed under the concept of dislocation. But *how* to turn that elusive flank—that is the question. The most obvious and direct way would be to march to that flank and attack it at an oblique angle. But obviously the enemy will oppose such measures, usually through his own countermaneuver.

In a symmetrical scenario, such as the European wars of the nineteenth and early twentieth centuries, the general most often found that *he could not use mobility to turn the enemy's flank.* He simply could not march fast enough to reliably find and attack that flank, because the enemy's speed in marching matched his own.

Therefore he used mass to compensate for lack of mobility. The idea here is twofold. First, a continuous line of soldiers will accomplish what a small, fast group of soldiers might have done: They will find and overlap the enemy flank. Second, we will prevent the enemy's countermarch, because the major portion of our army will remain engaged with the enemy line, thus holding them in place. In military parlance, this function is called "fixing."

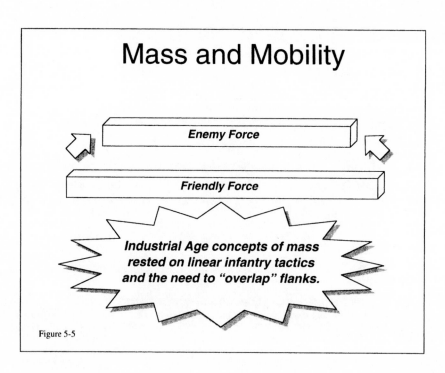

Figure 5-5

But what happens to the "overlap" idea when there is a mobility differential between the opposing armies? If we were in the shoes of the friendly general, what would it take to shake us loose from the goal of overlapping the flanks—a method of fighting that works only when we have superior numbers?

Would it suffice if we could move *twice* as fast as the enemy? Would that allow us to avoid the overlap dynamic and instead outmaneuver the enemy line? It is difficult to be certain, because of the various conditions that may pertain on any given battlefield. Moving twice as fast as the enemy probably would not suffice to consistently expose his flanks. But—what if our velocity were a *hundred* times greater than the enemy's? Clearly, if we possessed that great an advantage of mobility, we would immediately dispense with the wasteful idea of overlapping and instead race to the enemy's vulnerable spots and attack them, confident that he could not react in time.

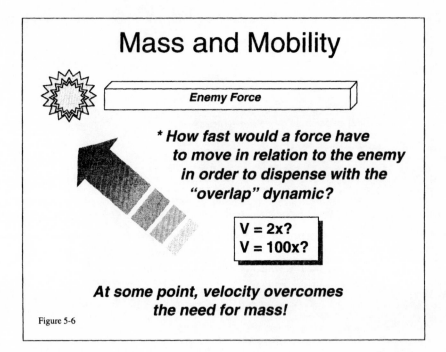

Mass and Mobility

Enemy Force

*** How fast would a force have
to move in relation to the enemy
in order to dispense with the
"overlap" dynamic?**

V = 2x?
V = 100x?

**At some point, velocity overcomes
the need for mass!**

Figure 5-6

We must conclude, then, that *at some point, a mobility advantage removes the need for mass.* The actual advantage required may lie somewhere between twofold and a hundredfold, but it is there. If we can demonstrably and reliably achieve this decisive mobility advantage, then one of the key arguments in favor of mass warfare collapses.

This is an important conclusion for the U.S. armed forces. As the army looks into the twenty-first century, its stated goals include a revolutionary leap ahead in both physical and mental agility. If we successfully attain it, then we must pull away from mass warfare ideas in order to fully exploit the mobility advantage. But there is a deeper logic still: We must forego mass *in order to attain mobility.* Just as in physics, mass trades off against acceleration—the larger the mass, the slower the rate of velocity change. If we rely on the principle of mass to build future forces, we will hamstring ourselves as we seek greater agility. It is time to progress from a slow, mass-based army to a fast, precision-based army.

Tomorrow's fighting forces will include small, lethal units moving with great velocity and precision to attack through weakness toward critical vulnerabilities. The linear tactics that the classic writers of the past wrote about are less relevant today, but the flanks and weaknesses they warned about still abound. Overmatching velocity—not overlapping mass—is the key to finding them.

Mass Equals Command and Control

Shortly after I assumed command of my mechanized infantry company in 1986, our battalion deployed to the field to train and conduct company-level evaluations. I was enthusiastic and anxious to maneuver my troops against the opposing force. On our very first mission, I was given the task of attacking a dug-in enemy platoon. As I analyzed the mission, I determined that our best chance of success was to fix the enemy frontally and maneuver around his right flank with an infantry platoon. Once the platoon was in position, we would mass fires upon the enemy to suppress him, and then assault with the flank platoon. Pretty standard stuff.

I wrote the operation order and briefed my platoon leaders. We rehearsed and prepared the soldiers. At the appointed time, we moved out and crossed the line of departure. I watched my flank platoon roll off into the woods with confidence and pride.

I never saw or heard from them again until the next day.

The attack was an abysmal failure. Somehow, the platoon leader of the flanking platoon got lost. He also managed to lose radio communications with me, with the result that he was completely out of the fight. With one-third of my combat power gone before the first shot was fired, things were not looking good. But they got worse. By the end of the mission, my company team had been slaughtered, and we had failed to defeat the enemy platoon. Rough start for a new company commander.

One of the lessons I carried away that day was not to let my platoons disperse too far from me. I had lost confidence in my lieutenants' ability to navigate and communicate, and I did not want to lose another platoon to lack of command and control. Fortunately, over the following two years, I got better at commanding, and my subordinates improved as well. By the fifth or sixth time we maneuvered, we had developed all the right standard operating procedures (SOPs) to overcome the inevitable navigation and communication problems. As our confidence in each other grew, our ability to maneuver effectively improved.

But I learned a trend that I have observed ever since: *Soldiers tend to concentrate in order to facilitate command and control.* When all else fails, it is easier to control your men when they are within visual range or, even better, shouting range. This is not just a fleeting observation based on one poorly trained company. It is a very old characteristic of tactics, and one of the tacit underpinnings of the principle of mass.

In small-unit tactics, leaders will often employ a "wedge" formation for movement. In the U.S. Army, a squad or fire team moves forward in a wedge with the leader at the front. The purpose of this wedge formation is to balance the need for movement with the need for firepower. If our only concern were rapid movement, we would not use a wedge; we would use a column formation, because it is the

Mass and Command
Concentrate to Control!

A *line* formation
for concentrating
fires . . .

A *wedge*
formation to
control fire
and movement.

"Follow me,
and do as I do!"

A *column* formation
for rapid movement . . .

Figure 5-7

fastest formation for movement. Conversely, if we simply wanted to concentrate our fires against an enemy to our front, we would use a line formation, rather than a wedge. But the wedge is useful, because it balances movement with firepower.

In order to make a wedge formation work, the leader moves forward, and his standing order to the other men is: "Follow me, and do as I do." If he walks forward, his soldiers walk forward. If he lies prone, they lie prone, and so on. Given the technological context of voice or radio communications and visual control, wedge formations are useful.

We use them at higher levels of command also. In the Gulf War, our entire brigade used a wedge formation. Our tanks, artillery, and infantry fighting vehicles advanced in battalions grouped together in a wedge formation. The flat, open desert of southern Iraq allowed the brigade commander to literally keep his eye on his entire command. Because fratricide and the orchestration of supporting artillery fires were serious concerns, the commander opted to keep the brigade relatively concentrated in order to more easily control it. In this, he operated according to the same paradigm as commanders in the ancient world. Keep the troops concentrated in order to keep them under control.

If we are to properly evaluate and revise the principle of mass, we must understand this aspect of the problem: Armies stay concentrated to facilitate command and control. But what happens to this equation if the technological context changes?

One of the most exciting technologies of twenty-first century warfare is "situational awareness." In the U.S. Army, we attain situational awareness through the use of a network of computers called "Tactical Internet." As of this writing, we have successfully built a Tactical Internet only in our heavy (i.e., mechanized) forces, but the future will also include light forces. The network consists of individual weapon systems—e.g., tanks, infantry fighting vehicles, self-propelled artillery, and even individual soldiers—equipped with tactical computers, position-location devices, and digital communications. Each weapon platform automatically communicates its position over the digital network, with the result that commanders and leaders can see the position of their subordinates and sister units on a computer screen. As the battle progresses, the commander can watch his unit move and fight on a digital map, and even track enemy units as well.

What does the Tactical Internet and situational awareness do to the principle of mass? It weakens it. When the commander and his subordinates can see each other digitally, they no longer need to maintain line-of-sight visual communications. When the technology works correctly—and initial tests showed it to be a very reliable system—it allows units to disperse more. When command and control does not depend on concentration, units can instead spread out and use the terrain more effectively. They can orient their maneuver on the enemy and terrain, rather than on the need to stay massed for control purposes.

Units that base their battlefield movements on the enemy and terrain, rather than on command and control, will give birth to a new

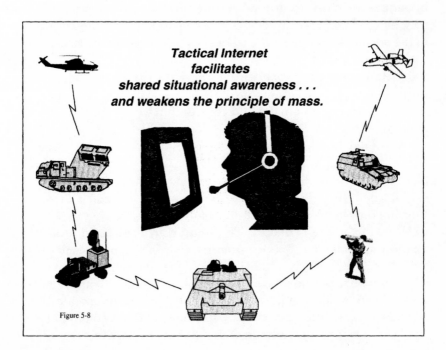

Tactical Internet facilitates shared situational awareness . . . and weakens the principle of mass.

Figure 5-8

kind of maneuver: *patternless* maneuver. Since the beginning of history, armies have fought in formations, and even the Industrial Revolution did not change this practice. Although armies dispersed and used terrain for cover and concealment, the requirements of command and control kept fighting units bound to formations. These formations constrained speed and lent themselves to easy templating by a knowledgeable enemy. But when digital technology frees us forever from the need for mass-based control, it allows us to come at the enemy in a patternless fashion that defies anticipation and complicates the enemy's understanding.

This technological trend is powerful. The implications exceed the imagination. But in order to exploit the potential, we must be willing to forego mass as a means of control. Just as mass fires give way to precision fires, so mass maneuver will be replaced by precision maneuver—facilitated by a dispersion controlled through digital communications.

Mass Equals Morale

Closely related to the question of command and control is the issue of morale. The pursuit of moral resilience in military forces under fire also impacts upon the principle of mass. Many historians have noted that, with the emergence of modern firepower and the consequent dispersion of men on the battlefield, the psychological phenomenon of "battlefield isolation" results: the numbing that occurs to men in battle who cannot see or hear their comrades. Isolated soldiers—often lying prone as protection from small-arms fire—cannot see or hear what is going on around them. They are susceptible to the feeling that they are alone and unobserved as the battle rages around them. In this condition, the tendency is to remain inert, not firing or moving. Whole units can ground to a halt due to this affliction.

Conversely, when soldiers in earlier ages went to war, most fighting formations put men shoulder to shoulder. Not only did tightly packed formations add killing power to units that fought with spear and sword, but they also conferred a moral strength that ultimately counted for more than weaponry. We know from countless studies that ancient battles featured little actual killing until one side or the other reached its moral breaking point. At this juncture, the broken unit would become disorganized, often turning and fleeing. Once that occurred, the real killing began, as the victors pursued and slashed at the backs of the routed enemy.

Commanders who faced this battle dynamic naturally did everything they could to increase the moral strength of their units. But the best way to maintain morale was to keep soldiers advancing together side by side. "Shoulder to shoulder, and bolder and bolder . . ." as the song goes. It was this fundamental battlefield sociology that was blown to bits by modern weaponry in the eighteenth and nineteenth centuries.

As modern armies adapted to the rapidly rising strength of modern firearms, soldiers dispersed to survive. But the resulting moral weakness remained a challenge down to our own day. Even when faced with the possibility of mass destruction, it is still tempting to keep soldiers and vehicles close together for moral effect.

What will the Information Age have to offer concerning this problem? One of the most crucial issues—yet one that has not been addressed at all—is the moral dimension of the digital force. Visions of future battle include a widely dispersed, patternless force of soldiers and vehicles tied together by a digital network. As we have seen previously, situational-awareness technology will facilitate the *control* of these dispersed forces, but what about their moral staying power? Can digitization overcome battlefield-isolation effects?

The jury is out on this question, but we must not avoid it for long. Army simulations will not provide the answer, because training simulations do not feature the pervasive fear of death and dismemberment that propel moral collapse in real war. Probably we will learn the hard way: in battle. But in anticipation of that future event, we must prepare through good training and systems engineering. We have enough experience and analysis to define the problem: Dispersed troops suffer from a lack of moral cohesion under fire. We must set ourselves to the question of whether digital technologies can help solve the problem. In the end, they must, because there are too many compelling reasons why we cannot revert to mass warfare. After all, a dead soldier has the weakest morale of all.

WHY MASS?

At this point, we have assembled classical reasoning behind the principle of mass and concluded that those reasons do not apply in the twenty-first century. We have seen that, due to advancements in weapons technology, modern armies have gotten up over the critical ratio of one man killing an average of one or more opponents, thus invalidating the proposition that mass equals killing power. We are heading toward unprecedented mobility in our future warfighting capabilities, thus removing the need for mass to find and overlap enemy flanks or other weaknesses. Likewise, the requirement for masses of soldiers or combat vehicles to compensate for uncertainty is an anachronism, because information technology is significantly improving our ability to see the battlefield with precision. That same technology is facilitating non-line-of-sight command and control to a degree never before imagined, removing mass as a means to enhance a commander's control of subordinates. Finally, we are looking to the technology of tomorrow to reinvigorate the moral strength of dispersed units, so that we do not have to rely upon mass for resilient morale.

Why mass? There are no valid reasons remaining. It is time to retire the principle of mass.

MASS AND THE OPERATIONAL PERSPECTIVE

The men who developed the principle of mass lived and fought in an age dominated by the concept of the decisive battle. They could not imagine a type of warfare centered around something other than the ultimate tactical clash between two massed armies. But the U.S. armed forces have a different perspective—one based on the operational level of war. And it is from this perspective that, as stated, we can regard the utter uselessness of mass as a principle to fight by.

Proponents of the principle of mass invariably look to the ultimate example of battle to prove their point. By observing the German debacle at the Marne, we can be instructed that Moltke the Younger failed to heed the principle of mass. But in some respects, this methodology for underpinning the idea is fallacious, for the advantages of mass come to light only in battle. When an army is conducting any of myriad other normal functions in war, mass is a hindrance. Hence, we can cling to the principle of mass only if we are willing to assume a juvenile perspective on war as comprising nothing but battles—with no movement, resupply, or subsequent control of population permitted.

This principle has never been able to govern all of warfare, but only that small portion in which actual fighting occurred. To instruct a neophyte (tragically, history has found some generals in this category) that he must maintain mass is inaccurate and totally misleading. There are *occasions* to mass, and others to disperse. The art is in discovering the most effective transitions.

We must allow that the current expression of mass in FM 100–5 calls for application of the principle "at the decisive place and time," hence obviating the arguments I have outlined. The problem with this statement, however, is the notion of "the decisive place and time." What exactly is this "decisive point"? If we still operated according to the Napoleonic paradigm, in which we glorified the *battle* as the centerpiece of our warfighting, then our concept of massing at the decisive point in time/space would pertain. But the theme of our warfighting theory since 1982 has been the discovery of "operational art." An operational perspective on modern war demands that we shift our focus from the *battle* to the *operation* or *campaign.* Whereas in the past, armies sought for decision in battle, today we seek decision in the larger context of the campaign. In this we are attempting to fulfill the ideals expressed in Sun-tzu's *Art of War,* in which the ancient philosopher taught that the height of skillful campaigning is to win before the fight begins. According to this logic, the battle itself is anticlimactic: The outcome is foreordained by our operational maneuvers.

It becomes apparent, then, that when the U.S. Army instructs us to mass at the decisive place and time, it is contemplating *battle* as the decisive part of warfare. In a sort of doctrinal schizophrenia, the army proclaims its freedom from the tyranny of the Napoleonic battle, and yet it clings to the very principle of mass that grew from that paradigm. Like a man who has forsworn his mistress and yet carries her photograph in his wallet, the army cannot shake itself free from Waterloo.

PRECISION WARFARE

If there is any science behind the idea of mass, it is the Lanchester equations. Lanchester worked on early aircraft design in England, and his mathematical models were designed to address the armament of the RAF's fighters, the Spitfire and the Hurricane. In brief, he developed two laws that became the mathematical underpinning of mass warfare. The linear law, which Lanchester applied to indirect-fire battles, stated that in order to have a two-to-one chance of winning, you had to have either twice as many weapons as the enemy, or your weapons had to be twice as good. The square law, which governed direct-fire battles (and incidentally became the basis for attrition models in the U.S. defense establishment), stated that a side's weapons had to be four times as numerous or four times as good in order to prevail.

The practical application of Lanchester's laws in the military engineering problem to which he set himself was highly successful. But the subsequent application to military theory in general has led to a widespread and sustained misconception that pervades our doctrine even today. Dr. James J. Schneider, in his book *The Structure of Strategic Revolution,* cited the works of Richard H. Peterson, Herbert Klemm Weiss, and Daniel J. Willard to show that the linear and square laws were "overthrown" and invalidated by rigorous analysis of battle from the mid-1860s onward. Lanchester's ideas on land warfare have been persistently refuted in subsequent studies, but since no model on warfare has replaced the simplistic (and alluring) mathematical logic of the square law, we have instead clung to a thoroughly discredited and disproved idea as the foundation of our doctrine and training.

Massing Effects

This harsh condemnation is easily dismissed through a clever semantic ruse within our written doctrine: the idea of massing *effects.* Modern proponents of the principle of mass have slowly come to the realization that mass doesn't work . . . and, indeed, leads only to greater friendly casualties. Even the densest observers of modern war have seen that putting soldiers shoulder to shoulder and charging them toward an enemy position simply doesn't work anymore. Early in this realization, we began to modify the old principle, which called for the physical concentration of forces, into a new version that spoke of a concentration of *fires*. The advent and development of gunpowder technologies replaced the point of the spear with a bullet and thus allowed soldiers to disperse (which they did ever so grudgingly over the years) while at the same time directing their fires at one spot. Hence, we could speak of massing fires rather than soldiers.

But the journey from the sublime to the idiotic came with the latest modification of the principle, which calls for the future warrior to "mass effects."

> Mass the effects of overwhelming combat power at the decisive place and time. (FM 100–5)

I reject the notion of massing effects primarily because it is too vague for application. It smacks of pedantry and seems designed to provide a fail-safe way of pointing to past conflicts and asserting: "He lost because he didn't mass effects . . ." whatever that means. Concerning future conflict, the principle as stated offers nothing.

The original phrasing of the principle of mass envisioned the physical concentration of soldiers as the most effective *means* of defeating the enemy. In many respects, this practice was valid and necessary. Even the idea of massing fires, in some respects, was meritorious in the context of early modern and Industrial Age warfare. But the latest idea of "massing effects" is not logical. The principle of mass traditionally recommended force concentration as a valid *means* that would create the desired outcome or effect. But by telling a future soldier to mass *effects*, our train of thought has jumped the tracks, for now our principle addresses the outcome directly. It no longer offers the means, and it speaks of the outcome only in a vague and not very useful manner. It is analogous to advising the general that "winning is good."

How does one "mass effects"? As a war planner, I have never seen this phenomenon in practice. Modern operational art instructs us that quite apart from concentrating our combat power in time and space, we must *distribute* it throughout a theater. The Napoleonic paradigm of a decisive battle is the conceptual underpinning for the principle of mass. But that paradigm has been thoroughly expunged both in theory and in the experience of warfare since 1815. Surely, if the "single decisive battle" conception of war has withered away, must not the resultant notion of mass likewise be retired?

We must make clear in our doctrine our intellectual break with the Napoleonic tradition: a tradition that cost hundreds of thousands of lives in World War I. But marking time by speaking of "massing effects" does not repudiate the old idea—it *reiterates* it. If the intent is simply to advise us of the importance of coordinating all of our arms and services, then why not say so directly and avoid the confusion of associating this idea with the old principle of massing soldiers on a battlefield? If skillful integration of all assets is the point of this modern interpretation of mass, then we ought to search for a new name for the principle and jettison the current appellation, because it has too much historical baggage. To warriors of the past, mass meant concentrating soldiers for battle, not integrating effects.

We may find, to our surprise, that the principle of mass is laid to rest by an unlikely actor: Lanchester. It was, after all, the Father of Attrition who first posited that to prevail in a direct fire fight, we must have weapons that were either four times as numerous—or four times as good. And perhaps it is the qualitative aspect of the square law that will emerge in the twenty-first century as the foundation for the theoretical antithesis of mass warfare: precision warfare. With weapons and other technology that have an extreme and sustained qualitative edge over the enemy, we can and must forego our ideas of concentration from the past.

IMPLICATIONS OF PRECISION WARFARE

Advancing to an understanding of precision warfare through the retirement of the principle of mass must not be merely an academic exercise. There are serious ramifications. If this principle is foundational to our current doctrine, then abandonment of the principle must result in fundamental revision of our doctrine. And if, as we claim, we are a "doctrine-based" army, then upon the revision of doctrine, all our notions of organization, training, leader development, and materiel must likewise be reexamined.

To illustrate the dramatic implications of removing mass as a principle of war, we need only look at the classical idea of how to "weight" a tactical attack. In the late nineteenth and early twentieth centuries, common tactical doctrine required a commander to designate a given unit or segment of his attack as the "main effort"—a term we still use today. The commander was then directed to "weight" that main effort—that is, reinforce and support it—with various means and measures in order to insure its success. One of the key methods of weighting the main attack was *to narrow its frontage*.

By reducing the width of the main attack's frontage, the commander could in effect increase the ratio of friendly rifles and bayonets to enemy soldiers in the main attack zone. Therefore, by narrowing the main attack zone, the commander increased the strength of the attack and its chances for success.

This idea is a clear manifestation of mass warfare, and by examining it in the light of today's technology, we can see how utterly irrelevant the principle of mass is today.

In modern operations, it is much more likely that commanders will *widen* the frontage of the main attack, rather than narrowing it. Because the means of destroying or defeating the enemy have grown beyond bayonet charges, the requirement to closely manage the ratio of opposing soldiers is not much of an issue any longer. In modern close battles, destruction comes from firepower, not muscle power.

Although there is therefore no longer a need to narrow the frontage of an attack, there are urgent reasons to widen it. Modern armies require road networks in order to move and sustain themselves. Even light infantry and airmobile units are dependent on roads to some degree. When planning a large-scale attack, one of the most basic problems is how to distribute a lot of units on existing road networks. If, in an attempt to stay true to classical mass warfare, the commander allocated a narrow frontage to the main attack, he would simply constrain his main effort from rapidly developing his combat formations, because they would be forced to use fewer roads than other parts of the attack. To weight a main attack in modern operations, give it lots of road space.

Widening the frontage of the main effort also allows the friendly units more options for effective maneuver. A narrow zone of attack can easily prevent a subordinate from finding and turning a vulnerable flank. Also, with the threat of modern weapons, a wide zone of attack allows the main effort to disperse its forces for better protection. Width allows a subordinate unit more freedom of action and creates opportunities for it to develop its combat power effectively.

The drawback to a wide frontage is, of course, the possibility that the main attack may then have to deal with more enemy units. In some cases, this factor may disallow a wide frontage. But most often it will not, provided that the commander's concept of operation allows the main effort to stay focused on a decisive task and purpose. Our tactical doctrine used to facilitate this better than it does now by distinguishing more clearly between an offensive "zone of action" and a "sector of defense." Although a zone and a sector look similar when graphically portrayed on a map, conceptually they were very different. In a defensive mission, tactical commanders were instructed by doctrine that they were responsible for every enemy unit that entered their sector. But in an attack, a commander was not required to deal with every enemy unit in his zone. The zone of action simply existed as a means of permitting maneuver, not establishing responsibility for enemy units. A good attack concept aims at destroying a *portion* of the enemy in order to cause a general defeat of the whole, rather than engaging the entire defense.

This simple example from small unit tactics shows that the principle of mass came about in a vastly different military context than we have today. Just as we no longer narrow an attacker's frontage, but instead do just the opposite to weight his attack, so also we must radically break free from mass-warfare ideas and move toward the opposite direction: precision warfare.

In the matter of organization, our current concepts of military force design are explicit manifestations of mass. Mass warfare requires large numbers of soldiers. Large numbers of soldiers require a simple hierarchical organization, using the "building-block" approach. For example, we group similarly trained and equipped soldiers together in squads; we then group similar squads together into platoons; similar platoons into companies, and so on. The major components of our corps are similar divisions; the major components of our divisions are similar brigades; the major components of our brigades are similar battalions.

This symmetrical building-block approach was certainly evident in the early modern period, and still pertained down to the mid-twentieth century. Today, we are seeing the breakup of that pattern. More and more, our tactical units are becoming less dependent upon similar, bulk components. Rather, they are evolving toward *multifunctional design*. This is a good trend and an expression of precision warfare. Eventually, when we can break with our past, we will cease designing units with "three of those," and instead design them with "one of this, one of that, one of the other" to achieve effective functional integration.

The retirement of mass will also free our tactical doctrine to more fully explore the potentials of dislocation. We must remember that, in the beginning, mass was simply a technique for winning in battle. In its infancy, mass was no more a principle of war than dancing is a principle of romance. It was and is simply a technique. Once we disassemble the principle of mass and ship it to the Museum of Military Oddities, we will be better able to explore the many other techniques for prevailing over an opponent, and dislocation is the strongest idea of all.

Simulations within the armed forces will suffer the most direct and dramatic impact of the death of mass warfare. Computer simulations, in particular, are so dependent upon mass ideas that they are wholly unsuited to simulate real war. There is an iterative effect here: The principle of mass underlies simulation design; simulations then prop up anachronistic mass ideas through their effect on training and leader development. A movement away from mass will have to include a renaissance of simulation.

Finally, the emergence of precision warfare in favor of mass warfare will change the way we develop and acquire materiel in the armed forces. The economical expression of the principle of mass is mass production. The Sherman tank of World War II—whose greatest characteristic was the numbers that were produced—is the best example of this paradigm.

But in the future, mass production of the implements of war will not work. Technological advancements happen too fast in the Information Age. Even the idea of "wartime production" is outdated, because future wars will be fought at a tempo and in a political con-

text that will disallow formal transitions to war either economically
or politically.

Instead, the future of materiel acquisition will be the rapid development and fielding of *prototypes*. The overwhelming numbers of
the Sherman tank will be displaced by the dislocating *quality* of tomorrow's weapons. The numbers produced will not be based on Lanchestrian attrition dynamics, but rather on the combined-arms warfare practiced by multifunctional fighting organizations. Production
lines will strive more for adaptability in retooling, rather than for
mass quantities.

Conclusion

The future of warfare is beckoning. The potential leap ahead in the
capability and relevance of future armed forces is ours for the taking. But to get there, we must drop mass warfare ideas, just as past
armies turned their backs on cavalry charges. We must not mark time
intellectually by speaking of "massing effects." Instead, we must declare with finality that we have overcome the limitations and inefficiencies of mass warfare, and that we are determined to leave it behind. Mass is dead. Long live precision!

6: Economy of Force

The rational distribution of force, this is our problem in war.
—J. F. C. Fuller

When you have it in contemplation to give battle, it is a general rule to collect all your strength and to leave none unemployed. One battalion sometimes decided the issue of the day.
—Napoleon

Money doesn't grow on trees.
—Unknown

Economy of Force: Employ all combat power available in the most effective way possible; allocate minimum essential combat power to secondary efforts.

The American bison has been called the most important animal in North America. Among the many traditions of the American Indian was the special relationship between man and bison—a relationship characterized by the respectful use of a precious resource. Certain tribes became adept at using the bison in most efficient ways: for meat, for clothing, for shelter. Even the horns and teeth were harvested and put to use. In fact, the Indian practice of making the most of a kill was a cause of great discord between the white man and the red. White hunters brought the great beast to the edge of extinction, and the ecological disaster was framed around a picture of stinking bison corpses littering the landscape, killed only for their hides while all else went to waste.

Wastefulness is an ugly part of man, and the practice of warfare is particularly vulnerable to it. The dynamics that dominate warfare—uncertainty, fear, error, miscalculation, and often incompetence—lead to uneconomical practices in war. At times, inefficiency leads only to time lost, treasure wasted, or equipment poorly used. But all too often, human blood is the price of ineptitude. Of all endeavors of humankind, warfare has the potential to be the most uneconomical.

Businesses in a capitalist society are guided by profit. In a healthy economy, competition among corporations leads each to strive for ever-increasing efficiency in production, distribution, advertising, and so on. Businesses must be economical, or they cease to exist.

Not so with armies. Uneconomical armies can thrive and attain new heights of inefficiency, so long as they can be propped up by force or by fear. In peacetime especially, there are no internal dynamics to the military art that lead naturally to economy. Staffs swell, resources are dispersed among low payoff activities, and expense grows. There are no profit margins to dictate efficiency.

This is not to say that armies never practice economy, for there are many examples of superlative efficiency both in war and peace. Economy *can* happen in military art; it simply doesn't happen naturally. The overriding need to avoid death while dealing it to the enemy tends to push considerations of economy aside. As a famous Prussian soldier, Moltke the Elder once said after witnessing a battle, "I have learned once again that one cannot be too strong on the field of battle." Moltke was wrong about that, but it didn't seem so at the time. Wastefulness in war does not preclude temporary success in battle, but it rules out sustained success in war. Economy of force—a most unnatural thing to think about when you're being shot at—is fundamental to conflict.

As a result, economy must be force-fed into military endeavors. If it is to be accomplished at all, economy must be directed, preached, and insisted upon. The most conspicuous evidence we have is the existence of a principle of war called "Economy of Force."

Why do we have principles? Principles serve to make us do things that we would otherwise not naturally do of our own accord. Principles are intended to *change behavior*. Therefore, when we see a principle of economy, we can deduce that the principle exists to change our uneconomic tendency.

Soldiering is, of all trades, the most uneconomical, and therefore the most needy of economy. In this chapter, we will see why economy of force is indeed central to the military art. In fact, without economy of force, *there is no art in warfare.*

Colonel P. P. Rawlins, MBE recently suggested that Economy of Effort (the British version of this principle) be eliminated, because it suffers from being too logical, too much common sense. Further, he claimed, it is dependent upon all the other principles of war and has no reality without them. In fact, the reverse is true: Economy is the independent variable upon which all other principles depend.

Economy of force is a valid principle, because warfare, like economics, is fundamentally the management of scarcity. J. F. C. Fuller concluded that economy of force is the central law upon which all other principles are based, though his reasoning was abstruse. Our route to the same conclusion is much simpler. We awaken in the morning and discover that we are not God, and that, therefore, our strength is limited. Several moments later, we discover the principle of economy of force.

If our strength were infinite—at any level of war—then warfare would simply be the mathematical and dispassionate distribution of strength until the objective of the war or operation is achieved. There would be no art in the business. But our strength is never infinite; our warfighting organizations are strewn with weaknesses and vulnerabilities. And where we do have strength, it is severely limited in time and space. Thus, the obligation to conduct the business of fighting economically. And thus, the *art* of war.

The essential wastefulness of warfare provides the backdrop against which economy must express itself. Warfare is primarily about destruction, and the human tools used in warfare are not reusable if inefficiently handled. Death has a permanence that makes war at once the most dramatic and the most uneconomical of human pursuits. War is played out with a pervasive ignorance on all sides, and ignorance breeds waste and imprecision in operations.

We therefore have a double need for economy in armed conflict: our very limited capability applied in an activity characterized by wastefulness.

Economy of force was first expressed simply as the converse of mass. That is, mass required the concentration of combat power in a single point in space and time. Logically, this concentration required a removal of combat power from other places, or at least a minimization of it. Economy in most areas facilitated mass at the critical point.

With the disappearance of mass warfare, we might suppose, then, that economy would logically diminish as well. If we view economy merely as the other side of the mass coin, then it must go away when mass expires. But economy, as we have suggested, is not linked necessarily with mass; it only appeared that way to classical writers, because they were so distracted by mass ideas. To someone fixated upon the need to concentrate for that one decisive battle, what else could be the point of economy if not to facilitate that ultimate event?

But in reality, economy is far more profound and far-reaching than mass ever was. Mass was a technique; economy is a philosophy based on human nature. Future warfare will not emphasize mass, but it will require a renewed emphasis upon economy.

Economy of force underlies all other principles, whether revised in this book or the traditional ones. Because we do not have adequate strength to oppose the enemy's strength continuously, economy of force calls for *dislocation* as an economic way of dealing with enemy capabilities. Because we do not have unlimited resources of time, men, or equipment (or for that matter, political will), economy of force insists that we orient ourselves to the *objective* and avoid needless expenditures that do not advance us to it. The principle of economy guides our efforts to *secure* our forces, reminding us not to strive for total security, but only enough to allow for the accomplishment of the mission. In like manner, the principle of economy of force pervades all our thinking about warfare—or it should.

The uniqueness of this particular principle is that it has no logical antithesis. We can look at the principle of mass and suggest the antithesis of dispersion. We can consider *maneuver* or *dislocation* and posit the antithesis of *fixing* or *confrontation*. *Reaction* provides the converse to *opportunity,* and so on. But there is no logical opposite to economy. It would be hard to assemble an argument in favor of Profligacy of Force or Dissipation of Effort. For this reason—the absence of antithesis—I believe economy should be elevated to a law of conflict. In Part 3 of this book, we will add two other laws to economy.

ECONOMY AND TRUTH

In his book, *The Foundations of the Science of War,* J. F. C. Fuller perceived a relationship between *truth* and *economy:*

> [W]ar has mainly been an instrument of waste, because of the ignorance of the soldier ... [T]ruth derives its power from economy of force, and trial and error, after endless experiment, arrive at truth by economizing force; perfect economy of force and truth are therefore synonymous.

Commentators on Fuller's work have criticized these passages as being inscrutable and overly philosophic. This is regrettable, because what Fuller wrote in the 1920s is about to come true with overwhelming impact. *Truth and economy are inextricably linked in war.*

Why is this relationship important to us? Because Information Age warfare—to the degree that it attains its theoretical goals—is all about *truth.* Doctrinaires and pedants shy from the word—like parishioners avoid the front pews—preferring to use terms less laden with moral overtones: information, intelligence, cybernetic dominance, and so on. But accuracy demands that what we are after is not simply information, because information may be false or irrelevant. What we want—what all commanders have urgently required in war—is truth.

Information Age warfare will naturally fall short of the imagination. We can conceive of a commander who knows every important detail about the battlefield around him: He knows where he is, where his subordinates are, and where the enemy is. He knows the true nature of the enemy—his strengths and weaknesses. He understands completely the external factors that will bear on the outcome of his fighting—political, cultural, environmental, and so on. Such a commander is easy to imagine, but next to impossible to produce. The reality of Information Age warriors will be something less than omniscience personified. But, we will make progress toward that goal.

The technologies that will rule the future battlefield will aim at bringing a far clearer picture of what is going on to the commander. Our materiel-development efforts are focused on reversing the trends in warfare since the French Revolution. In the nineteenth century, commanders became considerably more ignorant about the battlefield. The size of armies expanded beyond the effective control of a single commander. In the twentieth century, under the ineffable onslaught of machine guns and field artillery, troops dispersed, further complicating a clear understanding of the battlefield. The commander who once could survey his entire command through the end of a telescope now stood in a daze amidst fear, rumor, and ignorance. The battlefield of the Industrial Age suddenly appeared empty and confusing.

One of the casualties of war under these conditions was economy. Ignorance breeds miscalculation, which in turn leads to waste. Attacks are launched against vague estimates of where the enemy's weakness might be—estimates that are often wrong, with horrible consequences. Reserves are hoarded against the possibility of enemy attacks that never materialize, and battles are decided while troops that might have been useful are paralyzed by the ignorance of the commander. Military units carry huge, expensive stockpiles of supplies, because having supplies nearby is the only sure way of overcoming overburdened and often misdirected logistical support. Quartermasters, equally unaware of the true battlefield conditions as their commanders, moved supplies around with great inefficiency and waste.

Warfare, prior to the information revolution, was uneconomical in the extreme. But the vision of the future promises a change. Information technology will serve to redress the balance between ground truth and the commander's grasp of it. Commanders in the Information Age will get smarter. But with what result?

Warfare must become more economical—certainly in the tactical, and even in the strategic sense. Commanders who know more should fight better. That is the whole logic behind the information warfare concept. A smart missile is an economical missile. One missile flies with utter precision, freeing us from the need to send volleys of them. An army attacks with clear understanding of the terrain, weather, and enemy, obviating the need for masses of men to compensate for uncertainty. Truth gives rise to economy.

For this reason, warfare in the near future will feature an unprecedented application of economy of force. In some sense, we will return to the efficiency of ancient battle and even exceed it. What we *know* will utterly change what we *do*. As we have seen, all aspects of the military art are touched by information. Organizations will migrate from the inefficiency of mass toward the economy of functionality. Battlefield employment of forces will forego the use of blind, virtually immobile masses in favor of aware, agile units.

ECONOMICS AND WAR

The military art has much to learn from the science of economics. Both disciplines exist because of scarcity. Economics deals with the scarcity of resources, labor, and capital. The military art addresses scarcity of arms, combatants, and time. Both pursuits aim for maximum efficiency. Although the idea of economy of force is an old one, military institutions have not added rigor to the concept. In the field of economics, on the other hand, efficiency is a well-developed subject.

If we are to elevate economy to the lofty status of a law, we are obligated to study it further and graduate beyond generality into mechanics. If economy is a good thing, then how does a modern commander achieve it?

To begin with, let us borrow some concepts of efficiency from the study of economics: *allocative* efficiency and *productive* efficiency.

Allocative Efficiency

The essential problem in the military profession is not simply a scarcity of resources, but also the infinite demands that compete for those resources. There is always too much to do in war. A commander in the field today rapidly discovers that he has too much to do and too few resources.

As a result, the military art dictates that we must properly allocate our combat power to the most important demands and leave other

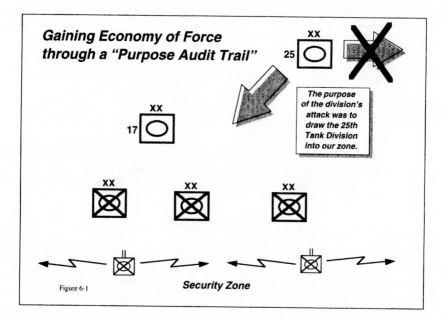

Figure 6-1

goals unresourced, or at least underresourced. In a sense, this is the classic notion of economy of force.

One of the greatest skills a general can own is that of knowing what *not* to do. This is allocative efficiency. In economics, it leads to the best use of resources and consequent high profit margin. In military operations, it conserves lives, morale, political will, and supplies. It involves refraining from expending resources on activities that are not important.

Mountaineers, according to George Leigh Mallory, climb a mountain "because it is there." But if that logic pertains to climbing, it bears no fruit in warfare. Attacking an enemy because he is there is a foolish and uneconomical expenditure of lives, time, and resources. The military art demands a greater wisdom.

One of the most practical ways of accomplishing allocative efficiency in the pursuit of economy of force is through the use of what we might call a "purpose audit trail." This is a planning technique in which the commander and his staff assign purposes to each component of the force, ensuring that those purposes are linked in a logical manner.

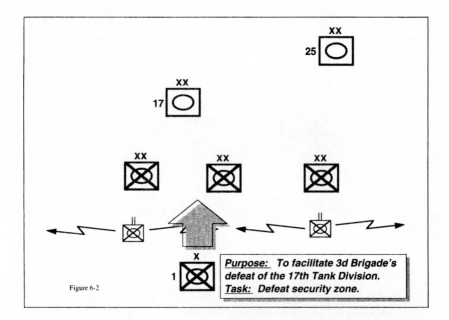

Figure 6-2

Purpose: To facilitate 3d Brigade's defeat of the 17th Tank Division.
Task: Defeat security zone.

I will use an example that happened to me during a warfighter exercise when I was serving as the chief of plans in an armor division.

The division's mission was to attack in zone to defeat the enemy first Combined Arms Army, for the purpose of protecting the western flank of our neighboring division, which was the corps commander's main effort. The key objective was to force the enemy to commit the 25th Tank Division into our division's zone of action, in order to keep it away from our sister division to the east. Our intelligence officer estimated that the enemy commander would commit the 25th Tank Division against us only if we defeated the 17th Tank Division, his second echelon defense.

The commanding general therefore approved a course of action aimed at defeating the 17th Tank Division in order to draw the 25th Tank Division against us. He first envisioned the end state of the plan: an attack by our 3d Brigade to defeat the 17th Tank Division. This would be the division commander's main effort. In order to most economically employ the rest of the division's combat power, the commander subordinated every other unit's purpose to serve the 3d Brigade.

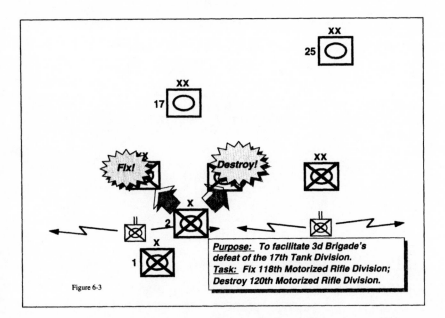

Figure 6-3

Purpose: To facilitate 3d Brigade's defeat of the 17th Tank Division.
Task: Fix 118th Motorized Rifle Division; Destroy 120th Motorized Rifle Division.

Ultimately, the plan looked like this: 1st Brigade would begin by attacking through the enemy's extensive security zone. The 1st Brigade's purpose was to facilitate the 3d Brigade's later advance and attack as described previously. Next, the 2d Brigade would conduct the attack to penetrate the enemy's main defensive belt. In order to blow a hole in the enemy's line, the 2d Brigade would fix (or immobilize) the one enemy motorized rifle division (MRD) and destroy the other. Again, the purpose of the 2d Brigade was to facilitate the 3d Brigade's attack of the 17th Tank Division. Finally, once the lead units had penetrated, the 3d Brigade would attack. The rest of the division's assets—attack aviation, artillery, engineers, logistical and signal support, and so on—were likewise aimed at supporting the 3d Brigade.

As in most operations, things did not turn out precisely as we planned. The attacks to defeat the security zone and to penetrate the enemy's main defensive belt were costlier than we had estimated. To complete the penetration, the commander had to make the difficult decision to allow the 3d Brigade to join the attack, thereby weakening the later main attack against the 17th Tank Division. We had also taken severe losses to our attack aviation. Finally, however,

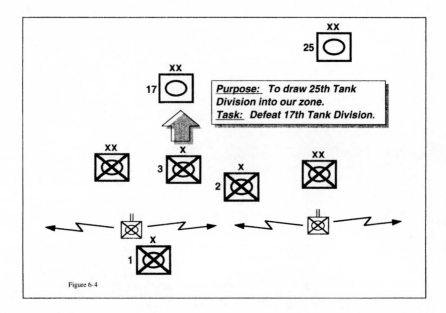

Figure 6-4

the division got into position to launch the 3d Brigade against the 17th Tank Division.

But just as the commanding general was formulating his last-minute instructions for the main attack, the intelligence officer confirmed for us that the enemy commander had already committed the 25th Tank Division against us. At first, we groaned. This meant that the 3d Brigade might have to face two divisions instead of just one, and our losses were already serious. But as we took a moment to consider the situation, we remembered the purpose audit trail that we had constructed. The purpose of our attack had been to protect our sister division's western flank by causing the 25th Tank Division to come against us. Even though we had taken serious losses, it became apparent to us that we had achieved our mission.

The division commander quickly grasped the essence of this sudden change in the situation. He called off the 3d Brigade's attack. Instead, he ordered the division into a tactical defense, allowing us to use a stronger form of combat against the enemy's advancing tanks. In the end, we prevailed, defeating the enemy's counterattacks.

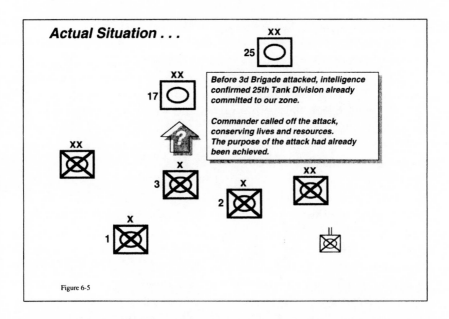

Figure 6-5

The genius displayed by the commander was in first formulating a *logical purpose* for each of the division's actions. Then, by considering the relevance of those purposes as the fight unfolded, he was able to rise above the confusion and call off what would have been a costly and utterly purposeless attack. We had almost attacked an enemy tank division just because it was there.

The use of tasks and purposes in military operations is a powerful part of army doctrine. It is also a key to economy of force in action. By staying true to the concept of allocative efficiency, the commander in war conserves his forces for the important activities. But there is another type of efficiency we must strive for also.

Productive Efficiency
The other form of efficiency within economics is *productive efficiency*. This is defined as combining available resources in such a way as to get the maximum output from consumed resources.

In the military art, however, we are arguably more concerned with *destruction* than *production*. We are about killing people and destroying weaponry. In practice, of course, armed forces also save lives, fight disease, safeguard property, and even deliver babies. But our core function—the part that makes us distinct from other governmental organizations—is our destructive capability.

Hence, we must consider *destructive efficiency:* our ability to destroy a maximum amount of the enemy with the least expenditure of resources (friendly soldiers, equipment, and time). Our goal, as we seek to economize in warfare, is to increase our destructive efficiency.

What is the most economical way to destroy something? We have already gained insights into this issue in the chapter on dislocation. Clearly, the best way to destroy something is by attacking through its weakness, rather than through its strength. It is easier to kick in a door than to demolish a wall. Because balance is essential as we contemplate future warfare, we know that sometimes we have to take on enemy strength directly to fix it or attrit it. But as often as we can, we want to go through the enemy's weak spots, because by doing so, we get the maximum payoff for the least expenditure of resources.

I have discussed combined-arms warfare in a previous chapter, so I will not repeat those points here. However, it is important to reiterate that warfare that pits like systems against one another in massive contests of strength on strength are uneconomical. The greatest expression of destructive efficiency in tactical operations is combined-arms warfare. The commander who comes at the enemy with a variety of threats and attack profiles can probe him, search for weakness, and attack that weakness before the enemy can adapt. Conversely, the commander who confronts the enemy with a single type of capability may find himself knocking his head against a brick wall if the enemy has developed appropriate defenses.

FORCE MULTIPLICATION AND COMBAT MULTIPLICATION

Closely associated with ideas of economy are the army terms "force multiplier" and "combat multiplier." Unfortunately, these terms are only loosely defined in our doctrine, and they are often used interchangeably. As we seek to improve our adherence to the principle of economy in the future, we must scrutinize and develop these ideas further.

Force multiplication is properly defined as *actions taken to permit one soldier to do the job of two or more soldiers.* Economy of force both in peacetime and war requires us to maximize the efficient use of our soldiers. We therefore use force multipliers to get the most from the limited numbers of trained soldiers available.

Examples of force multipliers abound. An automatic loading system on a tank permits a reduction in the size of the crew. A dual-purpose weapon such as the M203, which combines a rifle with a grenade launcher, allows us to perform two different combat actions with one man. Equipping and training a mechanic to repair numerous different systems allows us to reduce the size of logistical support units. In these and many other ways, modern technology provides force multiplication.

Combat multiplication is a different idea altogether, and it relates only to the battlefield, rather than to army operations in general. Combat multiplication occurs when *one weapon system causes an increase in the lethality of another weapon system.* As we have seen in our discussion of combined-arms warfare, the integration of unlike weapon systems causes the enemy to react in contradictory ways, thus making him vulnerable alternately to one system and then another.

Among the urgent tasks for the U.S. military establishment in the twenty-first century is the need to systematically and continuously develop force and combat multipliers as an expression of economy of force. The path to future success lies in developing ever greater destructive efficiency, and then employing allocative efficiency in action against the enemy.

Conclusion

Economy of force is not only a valid principle of war, it is the most important of those currently accepted within the U.S. armed forces. Because it transcends transitory methods based on technology, it remains relevant to future warfare and should be considered a law of war—one of the three that we shall develop in Part 3 of this book.

7: Objective

Victory is the main object in war.

—Sun-tzu

What does the politician want? Given this information, a general can base his plan on policy; but, if not given it, he must act on his own, and hope, against belief, that he will not be interfered with politically.

—J. F. C. Fuller

The ultimate military purpose of war is the destruction of the enemy's armed forces and will to fight.

—FM 100–5

Objective: Direct every military operation toward a clearly defined, decisive, and attainable objective.

As we take on the principle of Objective, we must be ready to challenge some of the most profound assumptions and beliefs we have about warfare. Within this chapter, we are going to break into what have become almost religious tenets within the American military establishment. We intend to smash some sacred icons here, and no doubt some feelings will be hurt. Nevertheless, if we can advance toward the truth about Information Age warfare, it will be worth it. To that end, I urge you, the reader, to suspend your attachment—emotional and intellectual—to the principle of objective . . . at least temporarily. Let's dig in!

As we have seen, the principles of war have normally been applied to all three levels of war: tactical, operational, and strategic. Such is the case also with objective. Just as we demand that our president articulate clear objectives for the strategic use of the armed forces, so also we instruct our lieutenants and sergeants to orient their tactical maneuvers on a clear objective. Of all the principles, the principle of objective has been the easiest to apply to all the levels of war.

In this chapter, we must look especially hard at the application of objective to the strategic level of conflict, for it is there that Americans have done the most violence to the principle. As we shall see, the whole idea of objective must be examined in the light of contemporary technology, but it is at the very highest levels of decision making that Americans have misapplied this principle. To correct our course and steer toward success in the twenty-first century, we must change the way we apply objective to our strategic formulation.

This is another of the principles that is deeply loved (to excess) within the U.S. armed forces, and the affinity has progressed to the grossest misconceptions concerning warfare in general and future warfare in particular. We shall first look at the epistemology of the principle and then investigate its often erroneous application. Finally, we shall see how recent developments have impacted upon the principle.

Objective is best understood as an expression—an outgrowth—of the principle of economy of force. As previously mentioned, the perpetual scarcity of resources obligates military leaders to economize in every dimension: men, time, supplies, and political will, to mention a few. Logically, this disciplined expenditure of resources leads us to focus our energies upon the aim of our operations and to withhold resources from secondary actions. We economize our efforts by fixing our attentions on the objective.

Historians have condemned military operations that wandered from the purpose too far. A classic example is Germany's campaign of 1941–42 in the Soviet Union. Hitler's vacillation concerning the objectives during the campaign caused his armies' strength to be dissipated as they reached alternately toward Moscow, Kiev, Leningrad, and Stalingrad.

In American history, the Vietnam War stands as the single most influential event in our understanding of objective. As our armed forces recovered from the strategic disasters of the 1960s and 1970s, it became apparent that we had no clear objective in Southeast Asia. Thousands of American fighting men lost their lives in vain, because the leadership was unskilled in deriving strategic objectives. As a result, our military doctrines now proclaim the great importance of the principle of objective, and our emotional attachment to it reflects the pain we felt from Vietnam.

Such heartfelt attachment to a strategic principle can be healthy. But in the American experience, the emotional commitment to objective can be harmful because, in the years since Vietnam, our interpretation of the principle of objective has gone astray. To see how we have migrated from the original principle, let us look briefly at past interpretations.

EPISTEMOLOGY OF OBJECTIVE

The principle of objective was a late-comer to the lists of military principles. The student of military history can read the works of Sun-tzu, Machiavelli, Frederick the Great, Jomini, and others without finding much emphasis upon this principle. What little the classics of military theory have to add to our understanding of objective comes more from inference than from direct assertion. If we were to try to find the earliest traces of objective, we might begin in 1589 with Justus Lipsius, a professor at the University of Leiden, whose *Politicorum libri six* proposed the idea that warfare should be rational and serve the interests of state. From Lipsius through Clausewitz and down to today, however, the principle of objective has been a minor actor. Why did the ancients not revere what some modern theorists believe is the most important of the principles of war?

First of all, the reader should by now understand that the principles of war, as they have evolved through the ages, began as principles of *battle*, not war. As citizens of a superpower, who think in global terms, the principles of war pertain mostly to our strategic thinking. But to the great writers, theorists, and soldiers of the past, these principles boiled down to winning battles on battlefields. As warfare changed and grew in terms of numbers of soldiers and technological complexity, the determinants of battlefield victory migrated off the battlefield and into campaigning, strategy, economy, and national policy. Battle would no longer be determined merely by the courage or virtue of individual heroes. The battle might still be the stage where the drama played out, but behind the curtain were an increasing number of team members involved in production, without which the play could not go on. The principles have had a difficult time keeping up with this complex movement. And when put to the test of rigorous critical analysis, they always reveal their tactical roots.

On the battlefield, in the midst of fighting, the principle of objective is about as relevant as a sutler's wagon. It's something we can attend to later. In the meantime, mass, maneuver, and surprise are much more important. We find that many of the great writers of the past have little or nothing to say about objective, because it is an idea that has little application to battle tactics, because in battle the objective is obvious: Kill the enemy!

Objective also suffered ignominy in the past, because of the political dynamics of "heroic" warfare. The heroic paradigm, simply stated, is this: Political issues will be decided based on the outcome of a decisive battle. This battle may unfold between two armies or even two individuals. But whatever the outcome, both political authorities and subject populations will acquiesce to the decision by arms.

War fought in accordance with the heroic idea tends to feature one or at most a few decisive battles. Because of this, the distinction between the *war* and the *battle* fades: War is viewed as pretty much synonymous with battle, and tactical victory becomes equal to strategic success. Thus, in the heroic paradigm, deriving the objective is simple: Destroy the enemy army in a single, decisive battle. There are no other issues or distractions.

Heroic warfare simplifies the transition from tactical victory to political success. When governments and people agree to abide by battlefield outcomes, the principle of objective fades in importance and becomes distorted. If it is mentioned or thought of at all, it becomes linked inextricably to tactical victory. Since the losers of the battle are presumed to agree to all the demands of the victor, the entire logic of the armed force is *to win on the battlefield*. The idea that soldiers would have to control terrain, compel obedience from civilians, or enforce economic policy is absent from the heroic paradigm. As Goliath proposed to the Israelites:

> Choose you a man for you, and let him come down to me.
> If he be able to fight with me, and to kill me, then will we be

your servants; but if I prevail against him, and kill him, then shall ye be our servants, and serve us. (I Sam. 17: 8,9)

In reality, there were many violations of this paradigm throughout history. Reluctance to accept a military defeat is as common as rebellion against authority and civil insurrection. But the idea of the heroic dynamic is long-lived and underlies the principles of war. For this reason, among others, the principle of objective was the redheaded stepchild of this ancient family. In the heroic tradition, objective simply was not an issue.

The final reason objective was not emphasized in ancient writings was that matters of state were not open for public consideration as a rule. The tradition of divine right of kings intruded into the utility and relevance of the principles of objective. But as kingdoms and empires slowly modernized and organized around economics, divine right was eventually replaced with conditional contract, and matters of state became less private. When the Roman senate or the British Parliament were required to provide money and men for warfare, they had an obligation to understand and consent to the objectives of the war. This concern with clarity of objective, of course, grew with the expense (in blood and treasure) of military endeavors. With that growing cost came the accretion of objective as a principle.

But, if objective was not emphasized in the original expressions of principles, how did it evolve? A careful review of past thinking will reveal two converging trends that gave rise to our modern notions of objective.

First of all, objective began life as the Siamese twin of mass. Mass warfare was all about concentrating forces toward a single point. Our modern idea of objective is precisely the same thing, except that it involves the concentration of efforts and energy, not the physical convergence of troops. Because of the essentially tactical outlook of earlier writers, the idea of objective was virtually indistinguishable from that of mass. If we were to resurrect Jomini and ask him, "What is the objective in war?" he would answer, "The objective in war is to mass." Objective was born as a tactical idea, closely linked to mass.

The other trend that gave rise to the modern principle of objective was the emergence of nation-states and total war. As Western Europe emerged into the early Industrial Age, and as warfare became a function of modern nation-states, the classical prescriptions about marching to a single point on the ground easily mutated into strategic and operational concepts of focusing energy on a single political outcome. It became important to give voice to objective, because it became possible to dissipate tactical energies. Whereas in the heroic paradigm the battle was the ultimate expression of strategy and national interest, soldiers and statesmen in the nineteenth century began to wrestle with the difficulty of making battle count for something. The principle of objective grew proportionately with the irrelevance and cost of battlefield victories.

In the Franco-Prussian War, the Battle of Sedan should have ended the war. Napoleon III, the sovereign of France, surrendered personally to the King of Prussia, and the encircled French army followed soon after. According to the heroic paradigm, the war was over. Instead, Paris erupted into insurrection, guerrilla armies took to the field, and the violence continued for months. Prussian thinking—and, by extension, contemporary military theory—now had to struggle with the complex problem of making victory relevant.

In the 1960s, American war leaders faced the same problem. In Vietnam, battlefield victories were easy to come by, but they became, seemingly, neutral factors—irrelevant to the outcome of the war. As we emerged from the miasma of Southeast Asia, Americans contemplated the principle of objective anew.

Hence, we have a paradox: The principle of objective was decidedly tactical in its origins, but it grew into prominence only when tactics could no longer decide a war.

The early expressions of objective that we find in British writing reveal these trends. By examining earlier interpretations of this principle, we can see how far the American version has fallen from the tree . . . for good or ill.

In 1920, British Field Service Regulations expressed objective thus:

> Maintenance of the objective. In every operation of war an objective is essential; without it there can be no definite plan

or coordination of effort. The ultimate military objective in war is the destruction of the enemy's forces on the battlefield, and this objective must always be held in view.

In order to correct the uneconomical practice of losing focus, the writer insisted on the need to "maintain the objective." In this principle's earlier expressions, the thrust of the thinking was not so much the *creation* or *selection* of the objective; but rather, the *maintenance* of it. In other words, the principle of the objective initially addressed not the objective itself, but rather the commander's orientation to it. In short, he was directed to stick to it.

In American usage, the first problems with objective began as the principle migrated from the maintenance idea toward the creation or selection of the objective. This was a critical if subtle change, because in modern warfighting the selection of objectives at the strategic level is primarily a civilian function, not a military one. In a democracy, we do not look to the generals to choose objectives in war, but rather to attain them. In the West, since before the early modern period, the principle of objective was subjected to a peculiar political/military context unknown to earlier rulers. On the one hand, there was a political authority responsible for the selection of the overall objectives in a war; on the other, there was a military command structure trained to believe (according to the tradition of ancient despots) that the selection of objectives was their business.

As the natural tension between these two entities developed, the problems related to objectives in war began to center around the *communication* of them from the political authority to the military command. Political authorities must deal with many factors—some of them legitimate, some of them indefensible—that prevent them from the efficient selection and communication of objectives in war. Some civilian officials, of course, are simply incompetent in such matters, either through lack of education or lack of aptitude. But more often, there are other factors that interfere. In the case of the United States, the power to declare strategic objectives has been shared between the executive and legislative branches. In practical experience, the selection of objectives connotes not only the ability to accurately develop the right answers, but also the critically important process of garnering public support and material or financial resources for them. What may be a perfectly reasonable and effective military objective may simultaneously be politically, economically, or culturally unacceptable.

On the military side, the desire for a clearly expressed, unchanging objective is intuitive. If only purely military factors were considered, we would seek to define objectives as early as possible and select objectives that were easily attainable given the means available. Further, we would insist on objectives that did not change within the course of the military operation.

In summary, we have factors that conspire to delay and obscure the objective, while at the same time we have the urgent need for an early and clear expression of it.

Within the context of this natural tension, the modern American application of the principle of objective has migrated somewhat from the original ideas. Still smarting from the Vietnam War, American strategists during the Reagan presidency insisted on a yet stronger application of the idea of objective. This renewed emphasis upon the strategic use of military forces was given voice in the so-called Weinberger Doctrine, named for former secretary of defense, Caspar Weinberger.

The Weinberger Doctrine, expressed in a speech in 1984, proposed that military force should not be used unless six tests were successfully passed: (1) the occasion is deemed vital to our national interest; (2) we should have the clear intention of winning; (3) we should have clearly defined political and military objectives; (4) the use of the military must be continually reassessed and adjusted as necessary; (5) the American people should fully support the use of the military; and (6) the use of the military should be a last resort.

There are, of course, problems with this doctrine. The last provision—that the armed forces should be used only as a last resort—is a particularly weak strategic theory. If practiced, it would guarantee that every scenario for the use of the military would begin with political failure—a most discouraging notion for military planners. To use the armed forces only as a last resort is analogous to a tactical system in which you attack with tanks and artillery but hold your infantry back until all other arms fail. Then, on the heels of disaster, you tell your infantry to advance.

A better approach is to *combine* all sources of power in order to achieve synergism. Just as in tactics we employ infantry, tanks, and artillery in a complementary manner, so also in strategy we should use the armed forces simultaneously with other sources of national power. Diplomacy, for example, is strengthened if, while the diplomats are speaking, we can demonstrate a military capability. The reverse is also true: Military force is much easier to apply when there has been successful political preparation. For example, if, through diplomacy, the United States has been able to gain access to critically needed ports and airfields, then that success reduces the requirement for a forcible entry into the theater by the armed forces.

The Weinberger Doctrine was not universally approved within the administration or the Department of Defense. Many professional military officers opposed the doctrine for one reason or another. And the armed forces were used in ensuing years without meeting all of those criteria. Still, the Weinberger Doctrine is a useful tool with which we can understand what happened to the principle of objective in the United States.

The requirement for clear objectives reiterated an important lesson from Vietnam, and almost all military officials agreed: The armed forces should have clear objectives in war. Or to be more specific, the National Command Authority should provide clear, stable (i.e., relatively unchanging), and achievable military objectives *before* the troops are committed. This idea has become a deified foundation to American military strategic thinking.

But let us examine these ideas. There are certainly points in favor of such demands on the part of the military. The early selection of objectives gives the National Command Authority and Congress the time to create political will and gather the resources and means to accomplish the objectives. Further, a timely selection of objectives gives the military establishment the time to plan (or more likely update existing plans), including the formulation of troop lists and deployment plans. The more time given to the military, the better they can develop effective courses of action, issue orders, and rehearse.

For all these reasons, it is logical to desire objectives that do not change. The changing of objectives threatens the military success that we desire in war. There is much practical as well as theoretical logic in the current military principle of objective.

There are, as we should suspect, also compelling arguments against these expressions of objective. As we saw earlier, we write to refute, and so we must deduce that our call for clear and stable objectives in war must be aimed against contrary ideas.

To begin with, the selection of effective and attainable objectives in war is a complex business. It is, in a sense, the grandest game of chance. The notion is that if I attain Objective A, then the enemy will capitulate and/or our national interest will be served. In the heroic paradigm, this causative linkage was not an issue. But outside of that paradigm, it is a most complicated problem. While simple in concept, the relationship between friendly action and enemy reaction is complex. The aerial bombing campaigns on both sides of World War II are clear examples of a failed assumption in this regard. Likewise, in Vietnam, we assumed that application of military power in various locations or against various entities would render a democratic solution in Southeast Asia. To our chagrin, such was not the case.

Secondly, things change in war. Who could have foreseen the remarkable success at Inchon in the Korean War? What began as an operation with the limited objective of retrieving the failures of the previous weeks quickly became a new strategic objective of totally defeating the North Koreans and perhaps reunifying the peninsula under Southern control. Unforeseen success caused the objective to migrate.

But more to the point for this discussion is the problem of tension between the general who wants to know the objective now and the president who isn't ready to decide. There are unreasonable extremes on both sides of this issue. On the one hand we have the all-too-familiar fiasco of Vietnam, in which the National Command Authority never developed and communicated a clear objective. On the other hand, there is a real (but perhaps unrecognized) demand by the military for total and immutable resolution of the objective before the start of an operation. Such a demand is easily defensible in theory, but it becomes, in effect, a demand for autonomy in action.

SUN-TZU AND THE AUTONOMY ARGUMENT

Sun-tzu is among the most famous of ancient Chinese military figures, and he is credited with leaving us his treatise on war, known variously as *Sun-tzu, Sun-tzu's Art of War,* or simply *The Art of War.* He lived in the fifth or sixth century B.C., and his book is required reading in most officer courses in the U.S. Army.

Probably the best-known anecdote about Sun-tzu's life and relations with his sovereign concerns the famous general's experiment in military drill using the king's concubines:

King of Wu: Can you conduct a minor experiment in control of the movement of troops?

Sun-tzu: I can.

King of Wu: Can you conduct this test using women?

Sun-tzu: Yes.

(The king thereupon agreed and sent from the palace one hundred and eighty beautiful women.

(Sun-tzu divided them into two companies and put the king's two favorite concubines in command. After instructing them in drill, Sun-tzu issued an order to face to the right. The women simply giggled.)

Sun-tzu: If regulations are not clear and orders not thoroughly explained, it is the commander's fault.

(He then repeated the instructions several times and commenced the experiment with another order to face to the right. Again the women laughed.)

Sun-tzu: If instructions are not clear and commands not explicit, it is the commander's fault. But when they have been made clear, and are not carried out in accordance with military law, it is a crime on the part of the officers.

(He then ordered the king's two favorites beheaded. The King of Wu intervened and canceled the orders, not wishing to lose his two beloved concubines.)

Sun-tzu: Your servant has already received your appointment as commander, and when the commander is at the head of the army, he need not accept all the sovereign's orders.

(He then beheaded the two women, and continued the experiment, in which the women all obeyed without a sound.)

No doubt this anecdote was intended to impress the reader with Sun-tzu's prowess and the importance of obedience in war. But there is a more subtle message in this story that relates to our subject, the American interpretation of the principle of objective.

Sun-tzu claimed that once the King of Wu issued his general the commission to prepare for war, and, to that end, to conduct the experiment on the women, the king's role and authority were temporarily neutralized as the general carried out the order. Sun-tzu expressed the notion that, although the military existed in subordination to the state, during the execution of an operation, the state sovereign had no authority to intervene.

Elements of this proposition are evident in many Western treatises on war. To simplify, Western military professionals have generally interpreted the relationship between the military and civilian authority as a sequential contract. *The armed forces exist to serve civilian authority. However, once the civilian authority has given the military a task to perform, further planning and execution is up to the military officials, and the civilian authority should not interfere.*

Moltke the Elder put it this way:

[A]t the moment of mobilization the political advisor should fall silent, and should take the lead again only when the strategist has informed the King, after the complete defeat of the enemy, that his task has been fulfilled.

Almost any citizen of a modern democracy would understand how absurd such a notion is today . . . and was even as it was spoken in Bismarck's Germany. Nevertheless, this idea of limiting civilian interference in military affairs has, in a watered-down form, persisted in American military thought. Underneath the veneer of humble compliance, there is a gut-level belief in most military men that war should be conducted without civilian interference.

But how would a soldier in today's American political culture express such an ideal? If the military establishment went to the president, the secretary of defense, Congress, or the American people with an insistence on noninterference in military matters, there would be an upheaval of reaction against military arrogance. Clearly modern American military officials cannot try the direct approach with this issue.

But there is another, more indirect way of obtaining operational autonomy: insisting upon an *objective*. The subtleties here are, perhaps, difficult to perceive. But in order to appreciate the depths of the principle of objective, we must understand that *insistence upon the laying down of wartime objectives is synonymous with a demand for autonomy in action.*

Let's illustrate with a simple analogy:

Most of us have played chess at one time or another. Let us consider the behavior of a chess player vis-à-vis his chess pieces. As he sits down to begin a game, the player will usually have some goal in mind. Obviously, the ultimate goal is to checkmate the opponent. But beyond that, a player will have some notion of how, specifically, he wants to approach the game. He may, for instance, have a particular opening in mind.

What does the player do with this general plan? Does he write it down and publish it so that his pawns can read the plan? Does he consult with the bishops and clarify his objectives for them? Does he provide a map with graphics for his rooks to use once the operation begins? Obviously not. The player keeps his counsel to himself. He does not commit himself to any particular plan, because he knows that opportunities will arise during the game, based on the interaction of his moves with the opponent's. Further, he is under no obligation to his pieces, because they make no decisions. The player makes all the decisions in the game. His will alone provides the impetus behind each move.

But suppose, for a moment, that chess were played with animated pieces that could think and decide for themselves. Suppose that the player, although nominally in charge of his pieces, were physically removed from the vicinity of the chessboard, and that he would receive updates of the game's progress only after every ten moves or so. What would the player have to do then?

He would have to delegate authority to the pieces. As the game unfolded, the pieces themselves would have to decide where to move, whom to capture, and so on—perhaps arranged according to some hierarchy (e.g., the king and queen supervise the pieces, while each minor piece supervises its associated pawn). In order to ensure unity of effort, the player would have to set forth some overall guidance: what opening to use, whether to castle queenside or kingside, and so forth. To be certain that the subordinate pieces accomplished his will, the player would have to describe the *objective*.

Obviously, our fictitious scenario does not pertain to real chess. But it does describe warfare under the conditions which prevailed in Europe during the past millennium or so. The head of state—king, emperor, or president—had to articulate a strategic aim of some sort, so that the generals and field marshals could act with relative autonomy in carrying out his orders.

Conversely, the officers charged with conducting the war learned to carefully elicit guidance from the government. In short, they learned to demand a clear objective. Not only did this help them to plan, prepare, and execute an effective military operation, but it also facilitated a tacit agreement by the government to allow the military a relatively free hand in attaining that objective.

In practice, of course, we can find numerous examples of governments interfering with military commanders. From the Duke of Marlborough to General Westmoreland, commanders throughout history have had to contend with civilian bosses peering over their shoulders. But the rule is proved in the breach: It was the specter of civilian interference that gave strength to the concept of objective. By securing an explanation of the strategic objective, the commander could in effect tell the civilian authority to leave him alone to achieve that objective his own way.

The point we must understand is that the principle of objective, as applied to the strategic level of war, carries with it a subtle, tacit demand by military leaders for autonomy in action. There are two problems with this demand:

First, it is unconstitutional. American political culture demands that the military arm be subordinate to the civil authority. This is an arrangement that has contributed to the remarkable success of the American experiment. There is not a soldier, sailor, airman, or marine who would openly challenge this subordination. But we must understand that an unbalanced demand for an early articulation of unchanging objectives comes dangerously close to an inappropriate shifting of authority from civil to military hands. In the end, if we fail to find balance, we may find our constitutional liberties facing the same fate as the aggrieved concubines of the King of Wu.

Second, there is another, more urgent problem with the unspoken demand for autonomy in action: *It won't work in the Information Age.* In the days of horses and sailing ships, there was a palpable communications gap between the battlefield and the throne. Kings and parliaments could not hope to supervise the detailed activities of their generals, because the former could not keep informed concerning the immediate details of the situation. As a result, the government issued broad guidance (i.e., the objective) and trusted military subordinates to accomplish the objective.

But today, the communications gap has all but disappeared. Governmental authorities have virtual presence on the battlefield, and this change in the communications context impacts upon the principle of objective . . . or it should.

This relationship between communications and objective is, perhaps, difficult to see at first. Let us use an everyday example from family life to illustrate the point.

THE SUPERMARKET ANALOGY

We begin by casting ourselves in the role of the family cook. We have an intention to cook a magnificent supper of chicken primavera tonight, but we currently lack the ingredients. Fortunately, we have adequate funds to purchase the necessary elements, and we have a teenage son to send to the supermarket.

Round One. We instruct our teenager that we want to have pasta tonight, and we slip a ten-dollar bill into his fist. Off he goes to the market, armed with enough cash, but an insufficient explanation of the desired end state. He returns an hour later with a can of precooked spaghetti. As we reprimand the youth for not bringing what we wanted, he reminds us of the principle of objective. "You should have explained exactly what you wanted." Hence, in Round One, we have illustrated a failure to communicate the objective in the context of the Agrarian and early Industrial Ages. Chastened, we try again.

Round Two. Having clearly explained in writing the ingredients we need for chicken primavera, we send teenage son back to the mar-

Figure 7-1

Figure 7-2

ket with the requisite cash. When he returns home, however, he reports that the required ingredients were not available. Therefore, he decided to act without specific authorization and use the cash to purchase another type of meal. In his mind, the overall objective was to create a supper for the family. He hands us a box of fried chicken, complete with baked beans and coleslaw. Frowning, we admit that he probably did his best. In this round, we have seen the classical application of objective at the strategic level of war, again in the context of the Agrarian or early Industrial Age.

Round Three. This time, before we send our teenager to the supermarket, we give him a cell phone. We have therefore introduced a new flexibility to the market scenario. Still smarting from his earlier reprimand, however, our teenager insists on a clear, written explanation of what he is supposed to do. We list the chicken primavera ingredients and send him on his way. About twenty minutes later, we receive a sputtering message on the phone that the market has no chicken. Disappointed, we direct our teenager to see if there is any shrimp, to which he replies in the affirmative. Hence, we adapt the strategic objective to the reality of available resources and change

Round Three: **Two-way communications add flexibility to the selection of the final objective.**

Figure 7-3

the menu to shrimp primavera. Two-way instantaneous communications have increased the success of the mission and illustrated the changes in managing strategic objectives in the Industrial Age.

Round Four. The next day, we send teenage son to the supermarket again, but this time we equip him with a cell phone and a helmet on which is mounted a real-time video camera. We push the money into his hands and tell him to head to the market. He demurs and insists on knowing the objective first. We explain to him that there is no need to resolve the objective yet, beyond the general notion of preparing a supper. He stamps his foot and points to the lessons of the past: The principle of objective demands that we state our goals before the mission begins and that we do not change them. Instead, we describe our desire for our teenager to go to the market, at which time we will be able to observe what is available. Thus presented with all the available options, we will choose more effectively and direct the teenager when we have de-

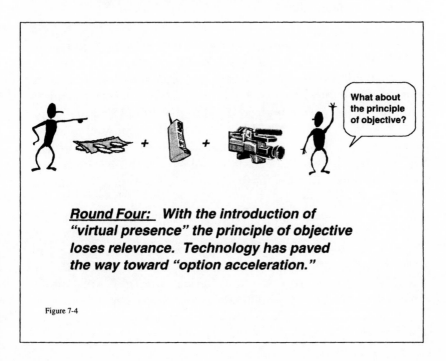

Round Four: **With the introduction of "virtual presence" the principle of objective loses relevance. Technology has paved the way toward "option acceleration."**

Figure 7-4

cided. This represents the communication of strategic objectives in the Information Age.

The principle of objective, as applied at the strategic level of war in the Information Age, must merge with what we can call "Option Acceleration." In the past, the head of state was physically separated from his generals and communications moved no faster than a horse or a sailing ship. In this context, the principle of objective had to allow for an effective, limited autonomy on the part of the general in the field. He had to know the intent of the government he served, but he had to have the authority to adapt the objective to the ground truth in the theater of war. As a result, the military profession correctly developed the principle of objective.

But in the Information Age, the head of state has virtual presence in the theater of war. In our nation today, the National Command Authority can literally watch a conflict scenario unfold as it happens. In fact, served with the full suite of information agencies and instruments available, the president and his staff could conceivably be more knowledgeable about the scenario as it evolves than the field general. Through the State Department, the president is in instant communication with other nations, including, most likely, the enemy. Through the intelligence agencies, he can collect human intelligence, satellite imagery, communications intercepts, and so forth. In the fast-paced modern world, the National Command Authority can sit atop a rapidly changing strategic situation that makes preliminary proclamations of strategic objectives mere guesses. Is it our desire to limit the flexibility available to the NCA through anachronistic, slavish obedience to a principle developed in the age of sail?

Instead, we must begin to investigate option acceleration: the rapid creation of political/military options in a theater of operations. The chaotic, often unpredictable context of both regional and global interaction will put option acceleration at a premium in future conflict. Twenty-first-century American strategists will plan for the use of military forces, diplomatic leverage, economic influence, informational resources, and intelligence operations to achieve national goals. The nation-state that can conceive, develop, and resource viable courses of action faster than the adversary will be the victors both in war and in lesser military operations. But to attain the best payoff from option acceleration, we must cease to overemphasize the misapplications of objective.

There are good and bad aspects to the principle of objective, as applied to the strategic level of war. We should embrace the lessons learned from Vietnam concerning the importance of feasibility and suitability of strategic objectives. But we should also exploit the opportunities that attend modern communications and information technologies. The modern regional commander in chief (CINC) must understand that with information technologies he can go beyond the old idea of focusing on a predetermined strategic objective and instead create a menu of strategic options for the National Command Authority. The measure of success in future contingencies may well be the pace at which regional CINCs can develop and attain such options. We can therefore balance the positive aspects of objective with the potentials of option acceleration to create a new and powerful strategic equation.

We use the term *acceleration,* because twenty-first-century strategy will require not just the *creation* of strategic options, but the *rapid* creation of them—at a pace faster than the enemy can match. Warfare is and will remain a time-competitive event, and future warfighters will be judged by how rapidly they can put viable strategic options in the hands of the National Command Authority.

It is common practice today to complain about "mission creep," a term that describes the migration of strategic objectives in a conflict. Military officials disdain mission creep and consider it to be a function of wishy-washy armchair strategists in Washington. Unfortunately, the next century will instruct us otherwise. Mission creep is *good!* It is an expression of option acceleration, and it is the theme of future military operations. Dyed-in-the-wool military conservatives will likely resist this trend, preferring to assume the fetal position, quoting anachronistic aphorisms about objective, and hoping that the future won't come knocking on their watch. The enlightened will instead acknowledge and adapt. Ancient warriors gloried in the numbers of enemies slain; future warriors will be measured by how fast they can deliver political options.

Option acceleration is no stranger to history. When Alexander crossed into Asia Minor in 334 B.C., the objective of the operation was merely to punish the Persians for their attacks on Greece and Macedonia. But victory at Granicus and Issus, and the gradual revelation that Alexander held the upper hand over Darius III, caused a migration of intent. Eventually Alexander conquered the Persian empire. Talk about mission creep!

Our vision of our strategic future should include an intimate and vital relationship between the political authority and the military instrument. The generals and admirals must be flexible and adaptive, comfortable with the demands of option acceleration. We must avoid the idiocy of LBJ poring over target lists on the White House lawn; but we must likewise repudiate the military arrogance of inflexible field marshals insisting that the minister of war keep his nose out of their business. War is about political intercourse, and it has no rationality when left to itself. When technology permits the political authority to direct the course of warfare more closely, we should not try to stay his hand by reciting dead, effete principles.

"LET'S GO KILL SOMEBODY!"
But there is also a more basic and conceptual flaw in the American understanding of this principle. One of the most erroneous ideas connected to the principle of objective within the U.S. Army is the notion that—in the words of the 1993 FM 100–5—"the ultimate aim of all military operations is the destruction of enemy armed forces." This assertion is totally false. Indeed, it is manifestly illogical and results in the army's desultory attempts to deal with "operations other than war."

The destruction of enemy armed forces is not the ultimate objective of military operations; it is the *preliminary* objective. The ultimate objective is the application of military force against nonmilitary objectives, such as economy, policy, or culture. This is a difficult concept for American military thinkers, taken as they are with the so-called interpreters of Napoleon—Jomini and Clausewitz. Both of these great thinkers focused on the destruction of the enemy (more so Jomini) as the centerpiece of strategy. Yet Clausewitz, in one of his most lucid passages, noted that in war the objective was political.

War is thus an act of force to compel our enemy to do our will . . . to impose our will on the enemy is its object. To secure that object we must render the enemy powerless; and that, in theory, is the true aim of warfare. That aim takes the place of the object, discarding it as something not actually part of war itself.

This is a profound passage from Clausewitz's *On War,* and in it lies both the dilemma of American thinking and the solution to that dilemma. Clausewitz noted that warfare does not exist for the sake of itself. Rather, the organized violence of war serves another purpose—Clausewitz used the "political" objective as typical of that purpose. But the purpose could be cultural, economic, religious, or otherwise, and Clausewitz's equation still holds true. Warfare always has a *purpose*—giving rise to the *objective* aspects of war.

But between us and the objective we seek stands an enemy armed force. Clausewitz thus noted that prior to attaining the objective we must first disarm the enemy—render him incapable of opposing us. This disarming of the enemy became the "aim" of war (as opposed to the "objective"), and Clausewitz noted that military men often focused on the "aim," setting aside the "objective" as not a part of war. Without knowing it, he was describing the dilemma of modern American doctrine.

The ultimate, nonmilitary objective *is* a part of war—just as important to the success of military endeavors as the preliminary conflicts with enemy armed forces. And it is by comprehending both the *objective* and the *aim* as part of war that we advance into a graduate-level comprehension of strategy, and we avoid the juvenile mistake of substituting the destruction of enemy armed force for the greater strategic purpose in war.

In this regard, army doctrine writers have made great progress. The most recent version of the capstone manual, FM 100–5, has dispensed with previous, unsuccessful attempts to define "low-intensity conflict (LIC)," or "operations other than war (OOTW)." Both of these terms were unfortunate, because they resulted in an artificial and ineffective distinction between the "aim" and the "object" in war, to use Clausewitz's terminology. As a result, LIC or its descendant

OOTW tended to be marginalized, both in doctrine and in resourcing.

Instead, the latest doctrine describes four categories of operations: offensive, defensive, stability, and support. The doctrine writers insisted that future military operations include all four categories. In a stroke of the pen, they solved the problem of the aim and the objective. By reminding future commanders that they must plan both for purely military operations (i.e., offensive and defensive actions) as well as for military/political actions (i.e., stability operations), the writers precluded an ineffective fixation on the former. This is powerful doctrine, but it remains to be seen how effectively the army and other services can assimilate it.

Part of the problem is the sequential way in which we still conceive the relationship between tactical and stability operations. Typically we view the big fight, followed by civil affairs. This is a paradigm of World War II. A more relevant example might be the conquests of Alexander. As the great ancient master maneuvered across the Near East, he constructed a campaign in which he combined successful tactical operations with far-reaching stability operations in a continuous cycle. Through good politics and wise governing, Alexander transformed small tactical victories into a sustained strategy. Our armed forces must come to understand—as Alexander did—that stability operations are vital components of the military art, not sideshows.

We shall dig further into the problem of "the aim and the objective" when we look into the law of duality. But to summarize this chapter, we must understand that the principle of objective must be influenced by changes in communications technology; if it does not, it becomes mystical dogma, rather than an expression of critical thought. We must also see that American interpretations of objective, especially since the Vietnam experience, have become unbalanced and come close to unacceptable demands for autonomy in military operations. Lastly, we must dismiss forever the absurd notion that the ultimate objective in war is destruction of the enemy.

It is the moral duty of military professionals to understand and practice *purposeful* warfare. When we adhere to old and utterly nonsensical ideas about objective in war, we are failing to fulfill that duty.

We must not permit the violence we create to become politically ir-relevant through an inflexible devotion to outdated notions about war. Just as the knight in shining armor eventually became the laughing stock of modern armies, so our military establishment can be emasculated if we fail to adapt to modern reality.

It is time to lay aside the old principle of objective. We can retain some of the wisdom that has grown from it, but the logic that un-derpins it does not pertain anymore. It's time to think about the fu-ture. Let's get out there and accelerate options!

8: Security

The whole art of war consists in a well-reasoned and extremely circumspect defensive, followed by rapid and audacious attack.

—Napoleon

Anciently the skillful warriors first made themselves invincible and awaited the enemy's moment of vulnerability.

—Sun-tzu

Security: Never permit the enemy to acquire unexpected advantage.

What would you do if I told you that next Tuesday, at precisely two-thirty in the afternoon, I was going to ring your front doorbell and, when you answered, punch you right in the nose? It would not be especially difficult to protect yourself, would it? Since you would know precisely where, when, and how I was going to attack, you could take any of a number of preventive steps. You could simply be elsewhere when I visited. You could refrain from answering the front door. You could alert the police and—provided they didn't think you were off your rocker—they might even await me and arrest me on the spot. In short, once you knew my intentions, it would require nothing short of suicidal incompetence on your part for my attack to succeed.

However, if instead of telegraphing my intentions, I simply made it clear that I don't like you and that I intend you harm, the problem becomes more difficult. Instead of the certainty of the Tuesday-afternoon-poke-in-the-snoot scenario, you have many more options to contemplate and protect yourself against. In fact, to fully secure yourself against what I might do would be expensive in the extreme.

Ultimately, you would have to move, change your name, alter your appearance, and take steps to secure yourself at all times. You would have to be ready for an unknown threat any time of the day or night. A very expensive proposition, and one that might force you to irrevocably change your entire lifestyle.

These two extremes define for us the military problem of security, and they provide us the framework to study an age-old principle. It is particularly appropriate for us to look at this issue, because information technology radically changes the nature of military security in future operations.

Security remains a valid principle of war in the Information Age, but it stands to be redefined. The problem with this principle is that its application is rooted in the cognitive darkness of *uncertainty*. Armies of the past, prior to the Information Age, were compelled to secure themselves not only from the enemy's action, but against the *possibility* of enemy action—against, that is to say, the uncertainty of war. In practical application, tomorrow's information-based armies will likewise have to deal with uncertainty, but to a far lesser degree. If the potentials of information technology are realized, the role of uncertainty in war will be lessened by a revolutionary measure.

When we can see the enemy with precision, clarity, and certainty, we economize our security. As we have noted previously, the law of economy demands that we minimize our expenditures of time, lives, and supplies in the business of security. The goal must ever be to secure the command as cheaply as possible. Resources wasted in security cannot be used for other activities. Ultimate security is equal to total inactivity.

In the past, when death and disaster stalked our encampments, clothed in darkness and doubt, we had to compensate for our ignorance of the truth with active security measures. Sentries, patrols, entrenchments, and so on, were the price of our profound lack of perspicacity concerning the enemy.

In the skirmish at Kelly's Ford during the Chancellorsville campaign in 1863, General Averell had a fleeting opportunity to destroy Confederate cavalry leader Gen. Fitzhugh Lee and his troopers. After forcing his way across the Rappahannock, Averell engaged Lee in a brief battle but was unable to gain a decision. In the end, the Union troops retreated.

Historians generally agree that the reason Averell failed to gain the day was because he had dispatched one-third of his command to secure against an imaginary threat to the North. As a result, they were not available to aid in the destruction of Lee's command, and the battle was lost. Averell, securing himself against the unknown, had sacrificed success.

In order to understand why the principle of security needs revision, we must understand that everything previously written about it was conceived in the context of the pervasive fog of war. When we read about security from Foch, Goltz, Marmont, or Clausewitz, we read persuasive arguments about the need for protection. These men wrote at a time when warfare was, arguably, at its "foggiest." The late nineteenth century had seen armies multiply in size and disperse over great distances like never before. The tremendous leap in lethality and tempo that modern weapons provided also combined to make the problem of security worse than it had ever been. Armies during this period thrashed about in the darkness of ignorance and had much to fear, much to secure against.

Warfare in the twenty-first century offers exciting new concepts of security—*if* we have the mental energy to adapt our tactics to the realities of technology. There is no end, however, to the skepticism that attends concepts on Information Age warfare. Soldiers and civilians alike voice their suspicions that our sensor technology may not deliver on its many promises, or that our staffs and commanders may succumb to cybernetic coma as they get inundated by information. To the degree that such questioning leads to healthy circumspection, it is good. But when it begins to become a roadblock to conceptual development, it is dangerous. Information operations are here to stay, and we must progress to making our investments in information technologies pay off. One way we can do this is to adapt our security measures so as to gain economies from our knowledge of the battlefield.

When we know where the enemy is, what his capabilities are, and we even have reliable indicators of his intentions, we can and must economize on security measures. Rather than reserve a large portion of our combat power for a hedge against uncertainty, we can more precisely tailor our security measures to meet only those feasible threats that are really possible and probable. Hence, information operations result in economical security. In the future, we will secure ourselves against the enemy, not against the unknown.

The direct result of information dominance over the enemy should be a relaxation of active security measures. This does *not* mean that we should naively drop our guard and invite destruction by a clever enemy. It means, rather, that the time, energy, and resources that we commit to security should decrease in proportion to the information that we have on the enemy.

The idea of information leading to relaxed security is nothing new; it simply hasn't been stated clearly in our doctrine because we have been used to relative uncertainty on the battlefield. But we have sufficient examples from the past to show us the way to building effective security concepts.

During the Battle of Britain in 1940, the British air strategists faced a dilemma: pilot fatigue. When faced with imminent, destructive bomber raids into their country, they had to consider the most effective way to counter those raids. The most obvious answer would be to fly constant fighter patrols near the English Channel, so as to detect and engage German bombers as they approached the English shores. The problem was that continuous fighter patrols wore out the pilots. With precious few aircraft and even fewer trained pilots, British air strategists had to conserve and employ their fighters with maximum efficiency.

The solution was to gain the information edge. To this end, the British constructed the Chain Home Defense radar system, which could detect inbound aircraft. A corps of human observers posted at key locations on the coast complemented the radar array, with the result that the Royal Air Force could detect, locate, and to some degree identify incoming raids. They learned to judge the types and quantities of aircraft and the likely targets they were flying to. In addition, the Allied breaking of the Enigma codes gave the British an information edge over the attackers.

What did the British do with this increased amount of knowledge? They relaxed their active security measures. With a fairly clear picture of the enemy, they did not need to rely as much on active patrolling by their fighter squadrons. Instead, those fighters could remain on standby on the ground and take off only when needed for an actual interception. Pilot fatigue was kept to a minimum, along with wear and tear on the aircraft. As a result, the British were able to conserve their most critical resource and prevail against the German air offensive.

Twenty-first-century land warfare must drive toward this same objective: to convert information into a more economical way of attaining security. Of all the measures taken by a commander to secure his unit, we find in retrospect that only a small percentage actually intercepted the enemy's blows. A somewhat larger portion of the security effort was directed against the enemy's *capabilities*. And, in our illustration, part of the resources used for security were utterly wasted, because the unknown threat guarded against exceeded the enemy's capabilities.

An unskillful commander, or one who suffers from the character defect of cowardice, will spend too much to secure himself. He will defend himself far beyond the enemy's ability to attack, thus wasting precious resources. When this happens, he will find himself short of combat power to fulfill his basic mission. He will run out of time, men, equipment, and supplies, but it is problematic whether he will recognize the fact that he is suffering from his own incompetence.

A better commander will not secure against the unfeasible. But he will take reasonable precautions not only against actual, detectable threats, but also against enemy capabilities. Since we have little control over what the enemy will choose to do to us, we must guard against what he *might* do.

Occasionally, however, our intelligence apparatus can determine not only enemy capabilities but also his intentions. Any good analyst knows that there are often discernible indicators of what an enemy commander is contemplating. For example, when he pushes supply stocks forward, it is generally a reliable indicator that he is intending to attack soon. Our efforts to develop the future force should aim at predicting his intentions as well as his capabilities.

I have heard and read the most ridiculous counters to the argument that information dominance should lead to a relaxation of security measures. It is frustrating and a little inconceivable, but there are senior officers in the armed forces today who believe that they can and must maintain a constant state of readiness and security within their commands. Admirable chutzpah, but intellectually bankrupt. As we will see from our study of surprise, military units are perpetually unready, which is why they require security measures in the first place. Further, while officers who boast of constant security come off well in peacetime, it is those same leaders who are notorious for doing nothing in war. Always secure means always inert.

Effective military leaders balance security and activity. They understand—sometimes intuitively, sometimes through study—that security trades off against velocity, and that velocity can indirectly accomplish security. The unit that moves and fights with overwhelming momentum eliminates the security problem by defeating the enemy. Of course, there are always practical limits to our ability to attack, so balance is required.

On the two extremes of this principle, we find a commander who can accomplish nothing because he has devoted all his energies to avoiding defeat, and another commander who loses his command because he did not take reasonable security measures against enemy capabilities. In the middle somewhere, we find reason and balance. But as information warfare provides an edge in the knowledge of the enemy, security must be economized, providing more resources for the main activity of the force.

SECURING OUR DIGITS

Probably the greatest amount of skepticism regarding the future of information warfare concerns the vulnerability of digital technologies to actions by the enemy. The fear of someone hacking into our battlefield computer networks, or jamming a key signal node, or spoofing our position-location devices is born of a healthy regard for the extreme complexity of digitization. A digitized division employs a vast array of computers, communications, sensors, and networks to achieve situational awareness and information dominance. With complexity comes vulnerability.

The argument that digitized forces are vulnerable is reasonable to the degree that it is based on fact. But there is a substantial amount of fiction that underlies some of these fears as well. Many of the skeptics—and typically the most vocal—know little about the realities of command and control warfare, but they have watched a lot of science-fiction movies and television shows. In the 1980s thriller *War Games,* we watched as a teenage hacker brought us to the brink of nuclear war with the Soviet Union. Our heroes from *Star Trek* blew up the formidable Borg Collective by feeding a faked command to self-destruct into their mysterious network. And in *Independence Day,* we were fortunate that the aliens who came to conquer the Earth had no computer virus protection. A single virus eliminated all their defenses. Spectacular!

I am convinced that these magnificent moments in science fiction have captured the attention—and to some degree the sanity—of many in the defense business. Vulnerabilities do exist in digital networks, but catastrophic collapse is hard to come by. In order to proceed with confidence and prudence, the army routinely aggresses against its own systems each time we train and experiment with a digital force. Specially equipped experts probe the Tactical Internet for openings into which they can insert viruses, or from which they can steal information. We continuously attack our own data, processors, networks, communications, and procedures in order to assess, identify, and fix weaknesses.

We have found, as common sense would predict, that digitized forces are somewhat vulnerable to command and control warfare. But when degradation results, it is typically limited in scope and effect, and the soldiers who were targeted learn quickly to protect themselves and adapt to the setbacks. This dynamic is true in all aspects of the force, and it is somewhat naive to imagine that digital technology would be any more vulnerable than any other.

Under the right conditions, it is easy to destroy an F-15 fighter. The M1 Abrams tank—known as the best tank in the world—is likewise vulnerable to destruction, given the right conditions. Every component of the force is subject to destruction or damage, and digital networks are no different. We know from our experience, however, that we can obtain a large measure of security against these vulnerabilities through thorough testing during development.

We will never obtain *total* protection, and we do not seek it. The principles of war teach us that total security is unfeasible and prohibitively expensive. What we want is *sufficient* protection—sufficient to allow our forces to accomplish the mission and defeat the enemy. This we can obtain for our digital networks. Attacks will occur and succeed against our digital forces. Degradation—sometimes serious, sometimes less so—will happen. But the advantages of digitization, if we exploit them, far outweigh the dangers and the costs.

Conclusion

In the past, military men wrote of the need for security. They lived in uncertainty and fog, and they were accustomed to protecting themselves against what they could not know. To them, such behavior was normal and fundamental to success in war. They could never have conceived of pervasive *knowledge* supplanting pervasive *fog*.

We live on the edge of the Information Age, a period that will usher in the greatest revolutions in warfare ever witnessed by man. The principle of security, as it stands, is not fit for duty in this context. It lacks balance, and it does not respond properly to the demands of the law of economy. In the future, concerns about security must be balanced with the need for activity, because we cannot accomplish our missions if we are overly securing ourselves.

Successful warriors in the twenty-first century will be masters of securing themselves with great economy, redeeming their scarce resources and converting them into activity, velocity, and victory.

9: Simplicity

Military plans soundly crafted will strive for . . . simplicity but rarely obtain it, because the several elements of any given force must be coordinated to carry out several different actions.

—Edward Luttwak

The power to command has never meant the power to be mysterious.

—Foch

Simplicity: Prepare clear, uncomplicated plans and concise orders to ensure thorough understanding.

Simplicity is not a valid principle of modern war. It is hard to imagine why this principle is still in our doctrine. Although there were valid reasons to consider this aspect of planning and fighting wars in the past, it is misleading today, especially when elevated to the position of a "principle."

The logic behind simplicity collapses in modern warfighting, but we will find in the end that, if we perform radical surgery on this principle, it will serve us well into the twenty-first century. Information Age warfare demands not *simplicity*, but rather *simplification*—a completely different idea.

But to begin with, let's first look at why the ancients called for simple plans. If they framed a principle calling for simplicity, we must infer that there were reasons that impelled commanders toward complexity instead.

Why would anyone deliberately make complicated plans? There are valid reasons for complex plans and orders. To begin with, a plan that is simple to your subordinates might also be simple to the enemy. If it is easy to grasp, it may likewise be easy to anticipate. Hence, even the firmest advocates of simplicity would recognize that there is a tradeoff between this principle and the principle of surprise.

But the other point is that in order to seek the advantage over the enemy and to obviate bloody attrition contests, we may need to develop plans that are inherently complicated. If an army consisted of one arm only—say, infantry for example—then the complicated business of coordinating between infantry and artillery would be avoided. But so also would we lose the synergy of combined-arms operations. The point is that complication is an unfortunate by-product of the search for advantage.

A large organization is more complicated to control than a small one, but we do not employ the principle of simplicity to reduce the number of soldiers. Rather, we live with the complication of a large army, because we know through experience that numbers can count (in some circumstances) in wartime operations. As we seek to dislocate the enemy, we employ numerous methods, arms, technologies, and maneuvers in order to gain the advantage. With each new component comes the need for integration and coordination—and complexity. Hence, in practice, we have always been willing to sacrifice simplicity for advantage.

The problem is that there is a point at which the complexity of an operation becomes a palpable obstacle to the friendly force. In the frustration that follows defeat, soldiers of the past have occasionally blamed failure on the complexity of the plan. Still, we are being less than truthful with ourselves when we state a principle that calls for simplicity without recognizing the bona fide existence of complexity.

It is unlikely that many readers would disagree with this point, but does this mean we should discard this principle? Certainly, there is an abundant and manifest common sense to the admonishment against overly complicated plans and orders. The intent of this principle, when balanced with other realistic aspects of war, is to encourage clarity of thought and expression.

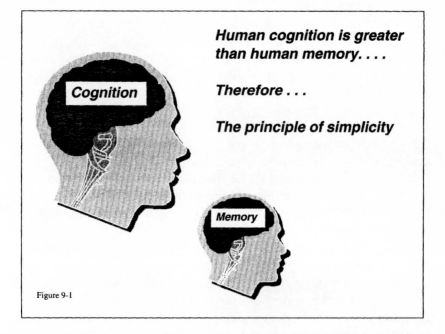

Human cognition is greater than human memory. . . .

Therefore . . .

The principle of simplicity

Cognition

Memory

Figure 9-1

The principle of simplicity has a theoretical, if tacit, foundation. The need for simplicity is based on the fact that human *cognition* is greater than human *memory*. That is, it is easier to think through details than it is to remember them under pressure. Hence, what cognition can create, memory cannot necessarily retain, and therefore, we discipline ourselves to make simple plans.

A logical train of thought, but is it as applicable today as it was, say, one hundred and fifty years ago?

In the days of Napoleon and Wellington, soldiers were often drawn from the lowest levels of society. In today's army, we attract recruits with promises of college education; in the past, promises of grog sufficed. Within some of the grandest armies of Europe were disease-ridden, alcoholic, illiterate hooligans, who sometimes took the queen's shilling to avoid prison. These brutish men were ill treated, ill fed, and largely despised by the society they protected. The solution to their technical training was endless drill. The key to employing them successfully in battle was . . . *simplicity.*

But is this the same type of soldier we are dealing with today? An anecdote from my most recent troop assignment will illustrate the point. During one of our gunnery exercises at Fort Hood, Texas, I was sitting in the tactical operations center, half listening to a private and a corporal arguing. The subject was particle theory. The corporal was finishing up a master's thesis on the subject! Shocked (and somewhat dismayed), I ordered them to stop this conversation and talk about beer, so that I could participate.

We live in a time when the social and technical context in which simplicity was developed into a principle has changed drastically. Today's soldiers (in the U.S. Army) are often surprisingly well educated, as well as masters of their military trade. Their ability to handle technical and tactical detail is far removed from the grenadier that fought in the Boer War.

Indeed, it is frightening to admit this, but recent technological advances have left us with a generation gap in the armed services. At the top military ranks, we have a generation that was raised and put on the uniform in a time before personal computers and digital communications. At the bottom of the chain of command, we have the Nintendo generation. When it comes to handling the technical complexity of warfighting, it is the younger generation that is most comfortable. In a sense, we have to keep things simple so we leaders can participate.

The computer revolution impacted squarely on the principle of simplicity—although we generally failed to notice. Someone once commented that the computer was invented just in time in human history: just as modern man's need for information processing threatened another Dark Age. Although there are realistic limitations to what the computer can do for us, it is increasingly an integral factor in modern military operations. Very soon, it will be impossible to find a soldier or a weapon system that is not equipped with a computer. Does this revolution in technology not impact on the principle of simplicity? It better! And it does. The question remains whether we can perceive that impact.

If the need for simplicity in military planning is based on the incapacity of human memory to process the complexity of human cognition, then the Computer Age must fundamentally change this principle. The computer has, in simplest terms, increased our memory. With the microchip, armies have gained not just the ability to store the text of an operation order, but rather the unsurpassed capability to store and process thousands of targets, to create and use logistical databases, and to manage personnel to an unprecedented degree. Add to this electronic distribution of orders and data, distributed rehearsal techniques, and myriad other potential applications of the computer, and it becomes easy to imagine the withering away of simplicity as a principle.

There are counterarguments to this proposition, however. The great unanswered question of the Age of the Computer is this: With the microchip, is man becoming smarter or more ignorant? So far in this essay, we have assumed the former. But there are those who portend a growing ignorance—an increasing gap between the data that a computer can *create* and the data that it can *process* and *store*. At issue is the ever-increasing amounts of data that computers linked with other computers can produce. On the one hand, we may be closing in on "all that is knowable." On the other, we may have invented a "breeder reactor" of information that has initiated a chain reaction, soon to propel us into a cybernetic coma of sorts.

If this latter interpretation of the Computer Age is true, then the principle of simplicity perhaps reasserts itself—not as the distasteful consequence of dealing with drunken recruits, but as the modern solution to cybernetic chaos. In either case, the principle of simplicity as we know it must be revalidated for the Information Age. If still applicable today, it must find its relevance in the *technological* context of modern armies, not the *human* context of the Victorian Age. In the end, the principle of simplicity might need a face-lift and a change of name: the principle of information management.

Whatever we do with this principle, we must learn to distinguish between the call for *simple* plans on the one hand, and *simplified* plans on the other. The former is a false and ineffective interpretation of the principle of simplicity; the latter is firmly grounded in logic.

Military planning not only contains inherent complexity, *it deliberately creates it*. See Figure 9–2.

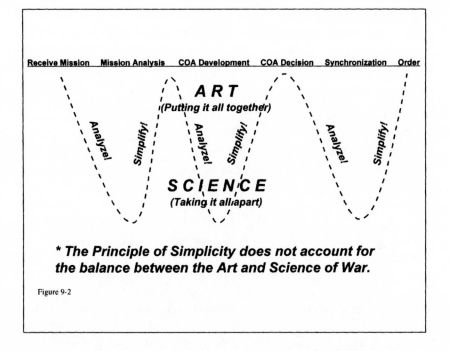

Figure 9-2

The U.S. Army's military decision-making model (modified to show how it is most often practiced in the field) shows how complexity and simplification interrelate as an expression of the art and science of war.

Science is a word whose roots connote cutting things into small pieces. "Scythe" and "scissors" are cognates of the word. In its most precise usage, the term "science of war" implies the analytical complexity that attends military decision making. As an example, modern army staffs, upon receipt of an order from a higher headquarters, will perform mission analysis—a rigorous and complex study of the mission, enemy, terrain, troops available, and time available in an upcoming operation. Likewise, once the commander has issued his planning guidance, the staffs must create and analyze various courses of action (COA)—again, a most complex and painstaking process. Finally, when the commander has chosen a particular course of action, planning staffs typically conduct extensive war-gaming intended to synchronize the myriad units and functions that must work together. All these steps demand complexity of thought and rigorous analysis.

The term *art*, on the other hand, has in view the "putting together" of pieces into a coherent whole. Art is creative, in that it brings together diverse parts into one integral body. Art is therefore the opposite of science. And, as shown in Figure 9–2, it also is a vital part of military decision making. For example, once the staff has conducted the mission analysis previously described, they must present their findings in a simplified format to the commander. That is, after finding out the most minuscule details of the operation, they must discriminate and set aside the irrelevant. They must find meaning in the facts, and communicate that meaning to the commander in a timely and easily understandable fashion. Following their detailed look at alternate courses of action, they must simplify their findings and make a combined staff recommendation on the *one* course of action they agree is the best. Finally, after their complex synchronization of the selected course of action, they must write an operation order and express the concept of operation as simply as possible to facilitate understanding and execution.

The point is that real warfare is not simple. Quite the reverse: It is almost inconceivably complex. The art and science of war demand a continuous process of analysis and simplification. We do not want simple plans; we want complex, effective plans that are simplified for execution.

INFORMATION MANAGEMENT

Not long ago, I was descending into Kansas City in a Boeing 707 when the pilot addressed the passengers. I awoke from an uncomfortable slumber to hear:

"Ladies and gentlemen, we will be landing to the west today on Runway Number Two."

Probably I was suffering indigestion, but this announcement irritated me. For the next six minutes, I watched the ground get closer and grumbled to myself: "Who the hell cares what direction we are landing?" I even went so far as to ask the pilot what possible difference our landing direction could be to anyone other than himself as he stood smiling and well-wishing the passengers on the way out. He seemed shocked by the question, so I decided not to lecture him about the wastefulness of disseminating superfluous information. But the point remains: There is a lot of information out there that doesn't matter. Nowhere is this more true than on the battlefield.

In 1997 the U.S. Army conducted two large-scale experiments in digitization. The first was a live, force-on-force experiment at the National Training Center, Fort Irwin, California. The second was a command post exercise at Fort Hood, Texas, using a computer war game to drive the action. In both experiments, we gained immeasurable insight into the business of information warfare. One of the lessons we learned was the need for information management.

There is a human side and a technological side to this issue of managing information. We anticipated the former, but the technical side has emerged as a much more concrete and difficult problem.

The human problem in Information Age warfare is the potential of what is often called "information overload." The idea is that soldiers can get inundated by battlefield information, much of which is either irrelevant or untimely. Even before serious experimentation began, we anticipated this problem and strove to create more efficient decision-making processes to mitigate it. The potential for information overload is a favorite topic among skeptics, and too much has been written about it. In practice, it is not such a serious problem, and we routinely solve it through good training and leader development.

The technical part of the problem is far more serious. Today's armed forces and the government agencies that support them can generate tons of information. The problem is that the available data has to be distributed through severely limited communications pipes. Bandwidth available for tactical communications is minuscule when compared with the information we want to move. Therefore, in a twenty-first-century application of simplicity, we have had to simplify our data in order to move it at the required speed.

What does this problem look like in battle? One example that we have seen is what limited bandwidth does to our tactical orders. Sending digital maps and graphics to subordinates requires a lot of time and large communications pipes. Early versions of the Tactical Internet simply could not handle the traffic. The data we tried to push overloaded both the communications backbone and the processing capacity of the computers. Systems lockups and crashes were frequent.

The solution is twofold: systems engineering and information management. Information technology will continue to close in on perfect efficiency in the communication and processing of data. But it will never reach perfection, and it is altogether likely that it will never even catch up with data requirements. Therefore, the need for active information management. It will be up to us humans to effectively manage the type and amount of information that we will distribute around the battlefield. In the early experimentation mentioned previously, units learned to reduce the size of orders and to simplify graphics. The solutions were not perfect, but they were good enough. The system lockups and crashes diminished in frequency.

But as we mature in information warfare, it will no longer suffice to manage information as an expedient. It will have to become a full-time part of warfighting. It has to become institutionalized, like the cleaning of weapons—a routine part of what we do in military operations. Modern staffs will have to include information managers, and every component of the military staff will have to be trained in effective economy of information.

One of the most obvious examples of failure in this regard is the army's fascination with tactical video teleconferencing. Not too many years ago, prominent general officers were insisting that they would never rely on any method to issue orders other than face-to-face briefings. They could not conceive of issuing orders to subordinates who were dispersed on the battlefield, because human sentiment and passion can only be communicated in person. (Those very generals were at the same time insisting that computers had no place on the battlefield.)

Less than a decade later, some of those same generals are leading proponents of digitization. This is good, because it shows we are a learning organization. But there is a holdover from the anachronistic "I gotta see your face!" school of thought: the video teleconference.

The armed services have used video teleconferencing for many years, and it is an effective tool, because it cuts down on traveling costs. Officers can meet over fiber-optic cables rather than flying to conferences. Almost all army posts have video teleconferencing capability today.

The problem is that we are trying to build the same capability into tactical units on the battlefield. Like Captain Picard on *Star Trek,* our commanders want to be able to talk to Star Fleet face to face, even while moving. Easy to do in Hollywood, but a very expensive proposition on the battlefield. Tactical video teleconferencing eats up huge chunks of bandwidth and may well require a separate communications architecture just to support it. It is a huge waste of money.

There are useful innovations in command and control for the future force. Among them is what is currently referred to as "collaborative white boarding." This is simply the capability to draw on a map or on a sketch board, while remote stations watch and participate. A commander, for example, could draw a sketch to explain a course of action in a battle. His subordinates and supporting arms could watch and also draw on the same map or sketch board as they build the plan together. A most useful tool for flexible command and control.

But wasting bandwidth and money to support the capability to look at each others' faces during tactical operations is a gross violation of proper information management. Modern generals who feel a need to look at their subordinates should carry a snapshot of them in their wallets. We do not need video teleconferencing on the battlefield. And if we are not skilled in deciding what we *don't* need, then budget cuts will decide for us.

The question of information management leads us to revisiting a fundamental argument concerning the future of command and control: the question of centralized and decentralized operations. To date, the argument has been dominated by bombast and traditionalism. Having studied the issue to some depth, I have concluded that the doctrinal fixation on the goodness of decentralized command and control has no basis in modern warfare. It is, in effect, a mythological principle. My brothers in the Marine Corps are the most afflicted by this delusion. Common wisdom in the Marine Corps is that maneuver warfare is inextricably linked with decentralized command and control. This is utter nonsense.

Fundamentally, there is no linkage at all between the form of command and maneuver warfare. True maneuver warfare is distinguished only by its foundational philosophy, which is dislocation. It is important to see that armies can achieve dislocation regardless of their command philosophy, because twenty-first-century warfare has the potential to permit centralization of command and control with powerful effect. To the degree that the Marine Corps and other organizations cling irrationally to the mythological belief in decentralization, they will miss the opportunities for maneuver warfare in the future.

Having said this, however, there is one aspect of modern warfighting that is a bona fide factor in favor of decentralization: bandwidth. If we had unlimited capacity to process and disseminate data, there is no doubt that Information Age warfare would feature a radical centralization of command and control. I have detailed this argument thoroughly in *Fighting by Minutes* and will not repeat it here. Centralization is the right way to go in modern warfare, *provided* that we can process and move data fast enough. But the real limitations in communications and data processing will obstruct centralization to some degree. As a result, there will be a viable need for a degree of decentralization in future warfighting. But to make it effective we must dismiss the religious fanaticism of current writings on maneuver warfare and come to grips with the realities of modern information technology.

Conclusion

The principle of simplicity is invalid and will remain so until we redefine it. In the past it was an expression of the difference between human cognition and human memory. It became an utter necessity when employing the infantry-based armies of the Victorian Age. But when applied to modern armies, many of which comprise highly educated soldiery, the principle lacks utility. Above all, the principle as stated fails to account for the fact that complexity is a by-product of the pursuit of advantage in war.

Nevertheless, there is a need for a modern expression of the balance between complexity and simplicity. The challenges are more technical than human, and they will serve as a constraint on the natural centralization of command that should result from greater information flow. In Part 4 of this book, we will subsume simplicity into other principles, but the logical and valid components of this old idea will continue.

10: Surprise

> Surprise should be regarded as the soul of every operation.
> It is the secret of victory and the key to success. It originates
> in the mind of man and accentuates the power of his will.
> —J. F. C. Fuller

> Rapidity is the essence of war; take advantage of the en-
> emy's unreadiness.
> —Sun-tzu

> At bottom, the strategy of surprise is nothing but an appli-
> cation of the principle of economy of force.
> —Erfurth

Surprise: Strike the enemy at a time or place or in a manner for
which he is unprepared.

Early on the morning of 9 December 1940, British Matilda tanks
from the 7th Royal Tank Regiment, supported by infantry and artillery, launched an assault against a fortified Italian encampment
along the Libya-Egyptian border, called Camp Nibeiwa. The attack
was part of the opening of Wavell's offensive into Italian-held Libya.
At the end of two hours' fighting, the British had captured Camp
Nibeiwa and 4,000 prisoners.

What is notable about this skirmish is the degree of surprise attained by the British. Although typical of desert engagements of that
period, Camp Nibeiwa shows the extent to which surprise charac-
terizes warfare. The British approach to the Libyan fortified belt went
undetected on the night of 7–8 December. Throughout the night of
8 December, however, the British shelled Nibeiwa and other forts.
Despite this fairly unambiguous warning that something was up, the
Italians were caught unprepared the next morning. A British feint
against the eastern perimeter of the camp confused the defenders

as to the direction from which the real main effort would come. (The British attacked from the west.) When the British launched their assault, they found an unprepared defending force: The Italian tank crews were dismounted and eating their breakfast. The British Matildas shot up twenty-five mostly uncrewed tanks as a prelude to the general assault on the Italian infantry positions.

This is a good example of tactical and operational surprise, but it serves also to demonstrate *technical* surprise. The Italian L3 and M11 tanks were no match for the British Matildas, even if they had been ready for combat. The Matilda represented a technical advantage for the British—one that could be overcome primarily through technical adaptation by the Axis.

Surprise is another principle that remains valid in the Information Age. In fact, as with economy, we will see this principle rise to ever greater prominence. Along with classic applications of tactical surprise, future conflict will feature technical surprise to a degree never before seen.

In order to fully exploit surprise dynamics, it is necessary to understand exactly what surprise is, why it works, and how it works. In this chapter, we will explore and demystify this basic concept of war that is often talked about but almost never explained.

Surprise is a battlefield condition that results from the interaction of two components: perpetual unreadiness and time. There can be no surprise in war without this interaction, so it is important for an army that wants to achieve surprise to understand these two factors and how they relate to each other.

PERPETUAL UNREADINESS

Imagination is a powerful deceiver. Much of what we think about warfare in general and future conflict in particular comes from the imagination. Sometimes, the images that emerge from our minds can be accurate and useful, but sometimes they are totally false. One such image that serves to warp our understanding of war is the fiction that our enemies are ready for battle.

Military units are perpetually *unready* to fight. Unreadiness is the natural condition of all forces, both friendly and enemy. Combatants in war are almost invariably oriented in the wrong direction, estimating the wrong threat, unsupplied, unrested, in bad terrain, ill informed, physically unfit, morally unprepared, or technically dislocated. Our mind's eye, conditioned as it is by our innate fears, fails to perceive this extreme condition of unreadiness, but it is there nonetheless.

Forces in war remain unready for combat for virtually the entire duration of the conflict, attaining a degree of readiness for only the briefest of moments before lapsing into unreadiness again. Even when they are ready, they are prepared only for a narrow band of threats. Any threat that emerges outside of that band will again cause unreadiness.

If it seems that I have overemphasized this remarkable and lamentable condition of military forces, it is because official doctrine and training within the armed forces does just the opposite: They perpetuate the myth that our enemies are always ready to fight. This is a dangerous and totally inaccurate view of the battlefield, and it is one reason why American tactical doctrine has never learned to capitalize on the principle of surprise. *There can be no surprise apart from the condition of unreadiness.* Trying to achieve surprise without understanding perpetual unreadiness is analogous to cooking a soufflé without admitting the existence of eggs. Unreadiness is the sine qua non of surprise.

Soldiers climb out of the condition of perpetual unreadiness when they perceive or anticipate a threat. This is why fighting forces throughout history have emphasized the importance of security. A typical combat force, if it has any skill at all, will, as its first priority, secure itself against unexpected threats. Whether moving, preparing defenses, or refitting, a good unit deploys part of its energies and resources to guard against the unknown.

Why do units do this? It is because they understand intuitively—but often not explicitly—that they suffer from perpetual unreadiness. Observation posts, combat patrols, battlefield sensors, and other security measures are all aimed at achieving a detection of a threat before that threat can come to bear against the force. Commanders use security measures to compensate for unreadiness through the gaining of the other ingredient of surprise: *time.*

TIME

It would be impossible to see a physical object that existed in only two dimensions. In fact, nothing can have physical reality without three dimensions. Likewise, the condition of surprise cannot exist apart from the dimension of time. Surprise is defined in *temporal* terms, not physical. Fish live in water; surprise lives in time. And it is by understanding the interplay between time and perpetual unreadiness that we can fully appreciate what surprise is and how to achieve it.

Here's how it works: At some point in the future, military forces will clash in a conflict. The term *conflict* here is intended in the most general sense. That is, a conflict could be a battle, campaign, or war. The conditions and factors that will bear on the outcome of that conflict are unknown and for the most part unknowable. Both sides will, of course, attempt to anticipate and control those conditions, but under the best of circumstances, all they can do is come close to the truth. When the conflict actually occurs, literally thousands of factors will interplay to determine the results. Good armies estimate those conditions well; bad armies do not.

Because the conditions of the future conflict are unknowable, both sides begin in a condition of unreadiness for that conflict. It is

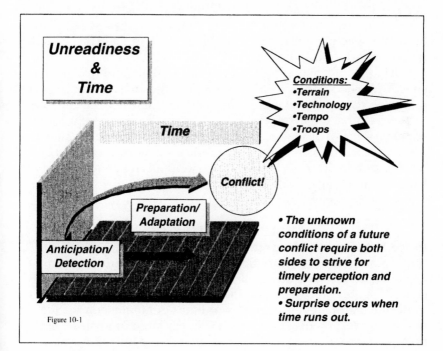

Figure 10-1

literally impossible to be prepared for the unknown. The question will be which of the two (or more) contestants will be able to prepare for and adapt to the conditions as they are revealed. Ultimately, both sides will adapt, but the side that does so first will prevail. The other side will suffer the effects of surprise.

The figure shows the processes of detection and adaptation occurring *before* the conflict occurs, because for any given force in battle, that is the desired sequence. As we know from history, however, detection and adaptation can occur *after* the conflict begins—or sometimes never at all.

If we overlay the concepts of opportunity and reaction onto this model, a bifurcation of competitive processes results. The *initiating* force *anticipates* and *prepares*. The *reacting* force *detects* and *adapts*. This is an important distinction, and it bears on the arguments concerning initiative in the Information Age that we have already discussed.

In the context of pervasive battlefield ignorance which dominated warfare prior to the Information Age, the best way to achieve surprise was through the use of initiative. This was because *anticipating and preparing* had a distinct time advantage over *detecting and adapting*. Therefore, the force that wanted to inflict surprise dynamics upon the enemy consistently would have to initiate the action, rather than respond to it.

But in the Information Age, we will find a subtle shifting of the balance. When knowledge replaces ignorance, the time advantage goes to the force that is *detecting and adapting*. Why is this so? Because *anticipation* of what might be is less accurate than *detection* of what is. And *preparation* for what might occur is less effective than *adaptation* to what is occurring. This by no means indicates a permanent and continuous shift in favor of reaction. But it does mean that classic aphorisms about initiative and surprise require careful revision. Ignorant armies must initiate. Aware armies can either initiate or respond.

TECHNICAL SURPRISE

In Joe Haldeman's classic science-fiction novel *The Forever War*, the humans and the aliens were fighting in the strange environment of time dilation. Because interstellar travel magnified the effects of this distortion, the combatants routinely encountered each other under wildly fluctuating technological conditions. A human combat force might depart for a distant star with the very latest in weapons technology, only to find upon their arrival that due to time dilation their technology was now hopelessly outdated. As a result, one of the first priorities in battle was to figure out the technological dynamics and adapt to them.

To a less dramatic degree, Information Age warfare will be characterized by this type of technological interplay. It is fundamental to our national security that we therefore prepare ourselves to compete technologically with future enemies. One of the most important aspects of that preparation is intellectual: We must understand the economic and political factors that will govern technological battlefield competition. Our goal will remain to achieve technological surprise.

What is technological surprise? It is not literal, sudden shock on the part of the enemy. We do not need to imagine future battle in which the enemy drops his weapon in awe and exclaims, "Gosh! I didn't expect *that!*" Surprise in the technical level of war simply means maintaining a general advantage and avoiding sudden disadvantage.

In the Information Age, nations that desire to have a sustained technological advantage will have to adapt both their economy and their warfighting doctrines to *prototype warfare*—a radically different approach to fighting than we are used to. In the Industrial Age, technology advantages were the result of generous investment and superior organization. Technological advances, because they occurred at a moderate tempo, fit into the paradigm of mass production rather easily. The rate of change, although extremely high when viewed from a broad historical perspective, was not high enough to overturn the way we ran factories and the economy as a whole.

But twenty-first-century warfare is infinitely more complex than, for example, World War II. Tomorrow's fighting forces are served by many thousands of technologies including not only hardware and software, but services as well. As long as the economy remains healthy, each of those technologies will continue to mature, causing not only a change in how a given technology works, but also a change in how the *system of systems* works. Therefore, future warfare will feature constant myriad technological advances that come at a tempo that disallows mass production.

The result is that a nation-state that opts for mass production will, at that same moment, drop out of the technology race. As their tanks, rifles, and bombs roll off the assembly line, they will be anachronisms. Conversely, the nation-state that wants to remain competitive will have to learn to think in new ways.

Prototype warfare requires a dynamic relationship among economy, organization, and training. Technological surprise is a powerful commodity—one worth fighting for. But to posture ourselves for success in this area in the future, we must shake ourselves free from Industrial Age paradigms that, as of this writing, permeate our doctrines and assumptions about war.

TACTICAL SURPRISE

I have written extensively concerning the dynamics of tactical surprise in a previous book, *Fighting by Minutes: Time and the Art of War*. To summarize: Although surprise is an ancient concept, we have not yet dissected it sufficiently in our doctrine and training. If we want to be true to this principle of war, we must become familiar and comfortable with the factors that create surprise.

We have seen that surprise results from the interaction of *perpetual unreadiness* with *time*. We have noted that when a force detects or anticipates a given threat, it takes action to prepare for or adapt to that threat. The goal of any force is to complete preparation—that is, to come to full battle readiness—before the battle begins. Given enough time to do so, a force will achieve this readiness.

Our objective, then, is to divest the enemy of the time he needs to get ready for any given conflict. At the tactical level of war, we deal with battles and skirmishes. In our pursuit of surprise, we will attempt to fight battles in which we are ready and the enemy is not. We achieve this effect by robbing the enemy of the time he needs to detect and adapt to a threat.

A bit of contemplation on this problem reveals two approaches we can use to take time away from the enemy: We can *delay detection*, or we can *hasten contact*. By delaying detection, we prevent the enemy from preparing/adapting by leaving him temporarily ignorant of the threat. Of course, the most dramatic example of this occurs when the enemy detects the threat only after the fight begins. In an ambush, for example, the enemy finds himself awash in sudden fires from unseen foes. The few enemy that survive detect the threat only after the engagement begins. They have a few seconds to adapt or die. This is why ambushes are so effective: They fight an unready enemy.

Most of the time, tactical surprise will fall short of the ambush dynamic. We can usually delay detection somewhat, but we normally cannot achieve continuous ambushes, particularly when we are attacking. Still, we can surprise the enemy by delaying his detection of us long enough so that he does not have enough time to come to full battle readiness.

The other side of tactical surprise is *hastening contact*. This simply works the other end of the equation, by rushing the enemy *after* he has detected us. The faster we can force an engagement upon him, the less time he has to come to full battle readiness. This is a simple concept, yet one that has profound implications for our future fighting forces.

Much of surprise is intuitive with soldiers. Throughout the ages, warriors have understood almost without thinking about it that moving fast is good—that it leads to advantages in battle. In fact, this need for speed is too intuitive, because what we know intuitively we fail to think about sufficiently. For this reason, we have collided with a doctrinal crisis concerning our future force. To wit: We are increasing the velocity and acceleration of our tactical forces, *but we have forgotten why we are doing so.*

One of the most basic themes of recent developments in Information Age warfare is to make our tactical units move faster. This we achieve by improving the speed of our decision making and by increasing our physical velocity. We are making great gains in these areas. The problem is that we are not simultaneously adapting our doctrine to the increased speed, with the result that our speed advantages are purposeless.

Armies increase speed for one reason: *to better achieve surprise*. If that is not the goal, then we should stop what we're doing and save our money. There are many soldiers and civilians who are involved in developing the future force that do not understand this basic purpose of speed. In fact, I have actually been lectured by a few diehards that the only purpose in going faster is to be able to get into attrition fights with the enemy at a more rapid pace.

This is not just nonsense, it is criminal nonsense. Racing into an attrition fight with a prepared enemy simply hastens the rate of our own destruction. It throws away all the advantages of velocity. It ignores the dynamics of perpetual unreadiness and time. We must, as an institution, discard this idea of racing into brick walls. For my part, if I am going to run my car into a wall, I am going to do it very slowly. I might, however, stomp down on the accelerator to get past a wall that isn't built yet. I use speed to preempt walls, not to crash into them. We must relearn the principle of surprise and remember that the only viable purpose of moving faster is to hasten contact with an unprepared enemy.

In *Fighting by Minutes* I showed that, in a context of symmetrical technologies, a fighting force hastens contact by employing *preemptive* tactics: tactics that sacrifice mass to gain time. The converse of preemption is *concentration* tactics: sacrificing time to gain mass. In symmetrical warfighting, the skilled commander learns to shift between these two forms of fighting—from preemption to concentration, and back to preemption, and so on. By alternately preempting and concentrating, he wins battles and then exploits those victories properly.

The key to victory in this context is balance, because a force cannot normally achieve preemption and concentration at the same time. They trade off against one another. But revolution in the art of war occurs when a technological advance allows us to both preempt and concentrate at the same time. The classic example is the invention of the stirrup. Before the stirrup, horsemen could not strike effectively from the saddle. Thus, cavalry were useful for pursuits and reconnaissance, but they could not prevail against heavily armed/armored infantry. With the invention of the stirrup, however, the horse soldier gained a stable platform from which he could employ the lance. The result was a revolution in warfare, a complete revision of the tactical dynamics of the ancient world. For the ensuing millennium, cavalry would dominate battle. European warfare saw the balance restored only after the invention of the longbow and gunpowder removed the shock advantage of the stirrup.

It is time to contemplate revolution once again. *We have invented the stirrup of the twenty-first century!* A recent technological development has again created the capability to both preempt and concentrate simultaneously. Unfortunately, the overwhelming impact of this technological progress has not yet been fully appreciated. The technology in question is *situational-awareness* technology.

As described in Chapter 1, situational awareness is the product of a Tactical Internet of weapons platforms tied together over an architecture of digital communications. Although this capability has existed for some time in the air and at sea, we have only recently demonstrated a reliable land version. But the thousands of technicians and soldiers who worked hard to achieve this capability have fallen short of recognizing its potential impact on our fighting doctrine.

What does situational awareness do? The stirrup gave striking power to a fast platform; situational awareness does just the opposite. It gives speed to a strong force. Tying a tactical land force together with digital communications increases the velocity of the force. Skeptics, and those who do not understand tactical warfare, are stymied by this assertion. Inevitably they ask how a force can move faster when the combat vehicles still move slowly.

Tactical velocity is determined by *formation*, not by vehicle speed. A modern battle tank can easily move forty miles per hour and even faster. But a tank battalion does not move even close to that speed on the battlefield. If a tactical unit—a brigade or battalion—moves ten miles per hour in operations, it is miraculous. Tactical units move slowly, because they move in formations.

In our chapter on mass, we saw that formations are related to the need for command and control. To the layman who has not tried it, controlling a battalion of combat vehicles seems an easy thing to do. In practice, it's damned hard. The speed that a battalion could achieve in the past was directly related to its training level. Fast battalions were those that practiced moving and communicating—a highly perishable skill.

But with situational awareness, tactical units move much faster—*potentially*. The reason they can do so is that they can answer the three fundamental questions that dictate movement in warfare:

- Where am I?
- Where are my buddies?
- Where is the enemy?

Because of the relative ignorance of Industrial Age armies, tactical units had to move in unwieldy formations that facilitated command and control, while at the same time hedging against unexpected contact with the enemy. But once we achieve situational awareness—once we know where we are and where the enemy is—the factors that slowed us down far below our vehicles' capabilities whither away. The result should be radically increased velocity.

Here's how it works: In the past, let's say a typical tank battalion commander wanted to move from one location to another location thirty kilometers away. If he could have been certain that the enemy was nowhere nearby, he could have lined up his tanks and other vehicles on the road in column formation and moved at maximum safe speed. He could have, in all likelihood, reached his objective in two hours. He could have gone even faster, were it not for considerations of safety and the control of his subordinates. The problem is that he did *not* know where the enemy was, and he could not hazard his command by risking a column formation—a formation that has virtually no striking power or protection. Therefore, the commander deployed at least part of his force off the road in a wedge, while at the same time deploying his air defense to cover his advance. He also took the time to thoroughly recon his route, and he assigned patrols to secure his flanks and rear. All this requires time. It would be several hours before the battalion closed on the final objective.

But what happens if, through situational awareness, the commander knows for certain where the enemy is? Likewise, he has a clear picture of his own forces. He can pass digital orders and graphics almost instantaneously over the communications network. In short, what happens to the speed of the unit with situational awareness?

It depends on whether they exploit their capabilities. So far, the outlook is discouraging. We have created the capability, but we have not exploited it. Like the nervous mother of a pilot who tells her son to "fly low and slow," we are hamstringing ourselves by not perceiving the advantages of speed. Situational awareness must lead to increased tactical velocity, and that velocity must lead to tactical surprise. So far, awash in the attrition mentality of past doctrine, experimental units have used situational awareness primarily as targeting data—with consequent ineffectiveness.

Situational awareness should result in a revolutionary application of the principle of surprise. It should convert into unit velocity, not just targeting data.

Instead, cautious doctrinaires are writing endless pages about how situational awareness can boost our lethality and protection. This is fine if not taken too far, but it distracts us from the greater truth: Our tactical combat forces are already lethal and well protected. But compared to how fast they *could* move, they are unspeakably slow. Once situational-awareness technology matures, we could have a tenfold increase in tactical velocity if we wanted it.

Why don't we seize this advantage? Because we do not, as an institution, understand the principle of surprise. We do not think about the perpetually unready enemy but instead conceive an enemy who is always ready for combat. It is dangerous to rush toward such a foe. Therefore, we languish at Industrial Age speeds, garnering combat power, massing our fires, wasting time.

In the end, the millions of dollars we spend on situational awareness return very little in real gains in lethality and protection, especially compared to the velocity gains we have failed to harvest. It is not too late. We can make this revolution in warfare occur, but the first step is to learn the real nature of surprise, and then build a doctrine based firmly upon it.

Conclusion

Some have written that surprise will be unachievable on the modern battlefield. The argument says that with information technology, competing forces will be able to see each other too well to allow any surprise. This is completely false. The twenty-first century will see the principle of surprise come to its fullest potential. Information warfare will see a constant battle between stealth and data fusion, between knowledge and ignorance, and between truth and deception. Surprise—which is, after all, a form of dislocation—will be fundamental to Information Age warfare.

The principle of surprise remains valid for the future of war. But to benefit from it, we must penetrate beyond mystical, intuitive understanding and get to the essential anatomy of surprise. We must come to grips with a fact that we have been afraid to look at: that forces in war are perpetually unready to fight. Once we understand that, we can use our technological tools and our tactics to manipulate time, so as to fight the enemy when he is unready.

The reader may be dismayed at the end of this book to find that surprise is not listed in the revised set of principles. If it is so important, then why isn't it still a principle? The reason is that surprise is subsumed into two other principles: the principle of *dislocation and confrontation,* and the principle of *distribution and concentration.* In order to fulfill the potentials of surprise, we must understand it for what it is: a subset of dislocation. While the term *surprise* is not listed in the last chapter, the idea is fundamental to the overall concept.

11: Unity of Command

Generally, management of many is the same as management of a few. It is a matter of organization.

—Sun-tzu

Nothing is more important in war than unity in command. When, therefore, you are carrying on hostilities against a single power only, you should have but one army acting on one line and led by one commander.

—Napoleon

There is one God that is Lord over the earth, and one Captain that is lord over the Pequod.

—Captain Ahab, Moby Dick, Herman Melville

Unity of Command: For every objective, seek unity of command and unity of effort.

He was awoken earlier than he expected that morning. The summer sky was a cool blue, but the sun had not yet begun to illuminate the dusky Italian countryside. La Trémoille, the commander of the French expedition into northern Italy, awoke from his dreams to hear the telltale clashes of metal and the screams of the wounded. A skirmish was joined, and it was alarmingly close to the house where La Trémoille lay.

It was 6 June 1513, and the French were about to be defeated by a numerically inferior force of Swiss pikemen. Before the morning was old, Louis XII's ambitions to recapture the Duchy of Milan lay in ruins.

The Battle of Novara is of only passing interest to the student of military history. Both the battle and the campaign illustrate certain aspects of the classical principles of war quite well—particularly surprise. La Trémoille's advance through the Alps preempted the Swiss occupation forces, with the result that the French rapidly conquered or cowed most of the Milanese towns. But the tables were turned on 6 June when the Swiss rapidly advanced to intercept the French investment of Novara. Against all expectations, the Swiss captains decided to attack from the march before all their forces had arrived. In the early morning hours of that day, the outnumbered pikemen— with no artillery and very little cavalry—crashed into the combined-arms French force and inflicted an embarrassing defeat.

As stated, the Novara battle does justice to the principle of surprise. But what does it do to the principle of unity of command? For the victors—the Swiss—employed what could be characterized as the most disunified command system in the history of Europe. In fact, they had no real command at all, despite the occasional use of a nominal *Hauptmann*. Their forces were led by practically independent mercenary captains, each controlling his own—and only his own— company. What they accomplished, they accomplished by consent and teamwork, without the benefit of a superior headquarters.

Hence, we have an indisputable example—it wasn't the first and won't be the last—that suggests that victory in war is not tied so inextricably to unity of command as the principle might suggest. Conversely, we could point to many more examples of strictly unified hierarchies that failed to achieve victory.

The principle of unity of command is too decrepit to be fit for modern war. This is another principle that has always been of questionable validity, but in the light of the realities of the twenty-first century, it starts to look like a suit of shining armor: very pretty and inspiring, but we'll leave it in the museum when we get alerted for war.

In reality, unity of command has always been nothing more than a technique for getting at what we really want: effective integration of battlefield activities. In its prime, this principle simply stated that unifying combatants under one commander was a proven and effective way of accomplishing this integration. A highly effective *method,* but not an end state.

Almost from the beginning of recorded history, this principle has run into trouble. Unification of combat forces under a single commander has a sort of rough-and-ready logic to it that has indeed scored many successes. Were it not for the fundamental importance of time in warfare, however, the very notion of leaving life-and-death decision making in the hands of one man would be ludicrous in the extreme: a most inefficient way of doing business. But warfare is bloody competition for minutes, and rapidity of decision making has almost always led to advantage.

The problem is that there are so many good reasons *not* to centralize authority. Since antiquity, military command has been viewed as a potential threat to the existing political order, and for good reason. A well-equipped, well-organized body of soldiers, loyal to their commander, can transform almost instantly into an armed rebellion and, in a few short days, into a bloody instrument of repression. In view of this fact of the history books, modern democracies, particularly in the West, go to great lengths to limit the powers of military commanders.

Even within the armed services, political realities tend to disunify command. As human beings, we suffer from greed, envy, ambition— even hatred, implacability, and vengeance. To suggest that the military art and the principle of unity of command is not touched by this fallen nature would be absurd. The right way to deal with the negative side of man's nature is not to try to eliminate it or ignore it, but rather to account for it in our theory, laws, and principles. Political activity and motivations will render ineffective any conclusions we arrive at that do not account for the essential humanity of war. But if we understand and assimilate human weakness, as well as human strength, into our reasoning, we will attain strong, effective warfighting concepts.

Apart from political reasons, unity of command weakened throughout the years as armies became both bigger and more technical. The human mind can handle only so much, and almost before military thought could be committed to writing, the art and science of war began to exceed the capacities of one commander.

As we have seen with other principles, the argument *for* unity of command grew in passion proportionate to the arguments *against* it. What the modern soldier and scholar must determine is the essential logic behind this principle (if there is any), and leave aside the traditions and assumptions that grew alongside it.

Almost everyone who has written about or thought about this principle in the last ten years has highlighted the problems with it. Almost all have come to the same conclusion: that the principle should be aimed at unity of *effort*, not unity of *command*. Unity of command is a means to the desired end of unity of effort. And while in the context of the Agrarian and Industrial Ages, unity of command was among the best means to achieve the integrated effects desired, that is no longer the case in the modern age.

The desire for unity of effort is again an expression of the law of economy of force. We cannot be diffuse and be victorious. The scarcity of our warfighting resources demands that we combine our efforts to succeed. When balanced with the requirements of modern distributed operations, the idea of unity of effort makes sense. The term itself—"unity of effort "—is, I believe, a weak one, primarily because it represents an effort to stay tied semantically to unity of command. What we really want is *effective integration*. No one who has viewed firsthand the chaos of armed forces operating with the State Department and private volunteer organizations would arrive at the term *unity*. But we can probably attain an effective integration—at least to a degree.

Unity of command, however, is losing ground in modern application. There have always been examples of successful, though disunited armies. The famous Swiss mercenaries of Renaissance Europe provide a good example, as shown previously. Rarely were their captains united under a single command, and yet they were able on many occasions to inflict serious defeats on their foes, because they combined their efforts. Likewise, the disunited high command in the Pacific theater of war during World War II produced synergistic campaigns that ultimately defeated the Japanese.

Unity of command, when exposed to the light of information technology, will begin to look more and more like the mystical notion of "divine right of kings" and less like a reasonable guideline for modern war. It is a frightening thought that the flow and distribution of information must, ultimately, undermine almost all traditional concepts of authority. I believe that the concept of authority will remain valid, but many of the trappings and traditions which surround it will have to go. If we critically examine the history and sociology surrounding authority in general and military authority in particular, we will find that knowledge and information flow are at the very heart of the whole business. Privacy has always been associated with authority: a symbolic way of protecting the right to information traditionally linked to authority. Exposing an authority figure to public scrutiny has always been tantamount to a radical revision of the hierarchy. From constitutional limits on monarchs to regicides to impeachment proceedings against presidents, information defines authority, and knowledge is the great kingmaker of human history. When information moves, it shakes up hierarchy wherever it is found.

For the purposes of this book, we shall confine our study to the implications of information technology to the problems of warfare in the twenty-first century and stay clear of the more philosophic aspects of information and authority. The bottom line for us is that information flow has changed radically, and as a consequence, our doctrines and traditions concerning authority and command are outdated.

FLATTENING THE HIERARCHY

Many soldiers and civilians have examined the problem of command in the twenty-first century, and their observations have been fruitful. This is an area that will experience some of the most immediate and conspicuous change as digital information technologies mature.

One of the ideas associated with modern information technology is that of "flattening the hierarchy." In the Information Age, the trend in successful businesses is to "flatten" the anachronistic and inflexible hierarchies that previously defined efficient commercial endeavors. The business world has produced some conspicuous success stories in which large corporations replaced rigid, bureaucratic organizations with a looser conglomeration of subordinate offices tied together by a vision of success. Likewise, in the complex art of joint, combined, and interagency warfare, the aged notion of a single commander overseeing all activity will not pertain.

But what does "flattening the hierarchy" mean? Phrases like this gain ascendancy primarily because they lack meaning. As modern, concerned citizens, we can envision a staid, inefficient bureaucracy in which workers have to get approval of three levels of authority before going to the latrine. Then we hear someone talking about flattening the hierarchy, and this seems like a great idea. But what does this flattening entail? Essentially, it requires the removal of intervening echelons of authority. Instead of one school principal in charge of six departments, each of which is composed of five teachers, each of whom controls thirty students, we have one principal in charge of 900 students. Prior to the Information Age, we would never have dreamed of such an arrangement. The argument we must rely upon in order to flatten the hierarchy is that we can compensate for the chaos of one-in-charge-of-many with technology, or with a transference of authority to the lower echelons, or both.

Flattening a hierarchy can lead to economy in operations, because, in the first order, we remove an expensive echelon of authority—hence, saving the cost of personnel, equipment, training, and so on. In the second order, we also hope to save time by obviating the inefficient referral of decision making upward and downward through various echelons.

As the armed forces continue to progress in information warfare, we will no doubt see continued scrutiny upon inefficient organization of staffs. Technology will serve to harness information and arrange it to be of use to commanders. But as the technology of war becomes ever more complex, military staffs will have to adapt—and sometimes grow—in order to accommodate it. In the end, technology will certainly affect the balance between large, complex bureaucracy on the one hand, and streamlined, small staffs on the other, but where that balance will be found is difficult to judge.

The other part of this solution is to revise procedures, so that there is a clear understanding of which levels of authority can decide each type of issue. Trips to the latrine are worker-level issues. Weekly work schedules are approved by the lowest level of management. Training requirements are handled at the next level, and so on. In this way, the organization avoids bringing innocuous matters to ever higher levels for resolution.

Flattening a hierarchy, we must remember, inevitably increases the span of control of the top echelons. To the degree that information flow and simplification of data can enable the top brass to handle that increased span of control, the flattening idea can bear fruit. In addition, some transfer of decision-making authority downward also allows for removal of superfluous echelons. But when complexity and lack of time combine to disable the commander, he must respond by organizing a staff to help out.

COMMAND AND ANARCHY

The more central question regarding twenty-first-century techniques for gaining integration of activities is that of *command* and *anarchy*. I have chosen these two terms, because they represent the most extreme end points of the argument between centralization and decentralization.

By "command" we refer to the ultimate hierarchical organization of subordinates under a legal, authoritative commander. By "anarchy"—the term means, literally, "without a leader"—we imply totally independent and unorganized component agencies striving to work together. Both concepts seek to accomplish the effective integration of battlefield activities, but one uses authority to do it, while the other uses consent.

The command approach to integration has advantages and disadvantages, and it requires certain enabling conditions to make it effective. The good part about a command-driven hierarchy is that it provides a simple, fast means for decision making. A commander can be liberal in his methods of controlling his subordinates. He can allow and even encourage debate and diverse opinions. He can loosen his grip and push decision-making authority downward. But when the pressure is on and time is of the essence, he can wave his hand and stop the debate. He can issue a decision and order it to be implemented. The history of warfare has taught us that a firm decision executed with alacrity often counts more than a well-considered course of action derived too late.

The disadvantage of the command approach is that it can prevent or constrain needed communication. It can serve actually to *delay* effective decision making, particularly if the commander has reserved too much authority for himself. At the Battle of Antietam in September 1862, the Federal troops at one point in the battle smashed the center of the Confederate line. For a few fleeting minutes, the Union soldiers stood looking at a huge gap that led directly to the Rebel rear—and ultimately to an opportunity to destroy the Army of Northern Virginia. But the command hierarchy failed as the decision whether to charge into the gap was referred from one level of command to the next. By the time General McClellan arrived to make the decision, it was too late—and more Union soldiers would have to die to retrieve the situation.

For command to operate effectively, the levels of *decision making* and *information flow* must be coordinate. That is, the movement of battlefield information should determine who makes what decisions. This is a simple and intuitive equation, but the essential truth of it has been lost in the mythological approach that the armed forces have always taken to command and control. A manifestation of this mythology is evident in the insistence of maneuver-warfare advocates upon "mission orders," "mission tactics," and the German concept of *Auftragstaktik*. Such proponents made the mistake of perceiving the success of the Wehrmacht in the early years of World War II, and concluding that the "mission tactics" approach of giving subordinates virtually all of the decision-making authority was a timeless principle of maneuver warfare. This is total nonsense.

Mission tactics worked (when it worked) because it adjusted authority to bring it in line with the state of information flow. *When* the tempo of information flow gives subordinates a more accurate and timely view of the battlefield, then they should have decision-making authority that is commensurate with that information. When, on the other hand, the higher headquarters has the information faster, decision-making authority should be centralized. Any insistence upon centralization or decentralization that does not emanate from this logic is mythological and not appropriate as a military doctrine.

Therefore, on the command side of this argument, we must have reliable and fast information flow. Information technology, as a rule, serves as an enabler for command. Part of the reason we have yet to see a clear exploitation of information technology is that we are clinging to outdated doctrine that calls for decentralized command and control. It is illogical, pointless, and a waste of money for us to make our higher headquarters smarter in future battle if we intend to fanatically preach the doctrine of decentralization. *Authority must follow information.*

The anarchy side of the argument also has advantages and disadvantages. The components of the friendly side of any future war will almost certainly include a large collection of players. The armed services—army, navy, air force, Marines, and, on occasion, the Coast Guard—will all participate in most theaters of war. Allied armed forces and government agencies will be there, each with their own agenda. The U.S. State Department, Central Intelligence Agency, Drug Enforcement Agency, and many other governmental organizations will be in theater. Private volunteer organizations and nongovernmental organizations such as the Red Cross, Doctors Without Borders, and others will participate. These players each have goals, procedures, laws, and institutional prejudices. But they also each have an enormous amount of expertise and competence within their own fields.

The anarchic approach is aimed at causing effective integration of these entities and their activities without constraining their effectiveness. Although a command-oriented hierarchical organization might streamline decision making, it would also militate against the productivity of these organizations. Competent sub-organizations need freedom and flexibility to operate, particularly in a context in which armed conflict is possible or ongoing.

To attempt to apply the aged principle of unity of command within this vast cast of characters is not only unrealistic, it is illegal. Constitutional limitations on the authority of the armed forces prevents unification of these organizations under military command. In a sense, this is a shame. There is no organization in the world better equipped and trained for leadership than the U.S. armed forces. The men and women in uniform are trained from their reception into the service to take charge and be decisive. The leadership skills that are so often wanting within other agencies of the government and civilian organizations are available in abundance in the armed services.

But the wisdom of the Founding Fathers was that giving too much authority to military commanders is a threat to freedom. The Constitution prevents it. Hence, at the theater level of war planning, it is anachronistic to speak of unity of command as a viable principle. Instead, the theater commander must be adept at orchestrating anarchic groups.

The art of orchestration requires many skills not traditionally found or emphasized in the military: diplomacy, patience, consensus building, and imagination. Of these, the last is the hardest to cultivate. Orchestration requires imagination, because, just as the conductor of a symphony hears the music in his head—hears what it *ought* to sound like—and waves his baton accordingly, so the military commander must be able to imagine what the end state of his forces' efforts should be. He must appreciate, even before it happens, what sister components can do for him. He must be able to perceive that even unwilling, uncooperative partners can contribute to his success.

But if orchestration of anarchy is required at the strategic level of war, what about at the tactical level? Won't future warfare still feature battlefields in which commanders must issue orders rapidly and decisively? No doubt they will. But the degree to which strict hierarchy will apply must ultimately depend upon information distribution.

Even in today's armed forces, unity of command is hard to come by. As an infantry battalion operations officer, this was very clear to me. According to the principle of unity of command, the battalion commander should have authority and control over the activities of his battalion. He should be able to unify the actions of all subordinate elements. That's the theory, but in practice, there are countless violations. The supporting field artillery, including the fire-support teams that are subordinate to the battalion commander, responds not to the commander, but to the fire-support officer in the brigade staff, and ultimately to the division fire-support officer. In my experience, the priorities of these different echelons of command are more often in conflict than consonance. As a result, I have fought many battles—both in real war and in training—in which the fire support was completely irrelevant to the battalion's combat power. This is not a condemnation: The reason we fight like that is to economize the use of the artillery. The point is simply that unity of command does not pertain even at small-unit level. Instead, the commander must attempt to orchestrate the artillery support, for he will never command it.

Likewise, the logistical support of the battalion answers to several bosses. In fact, future army organizational innovations will likely see logistical support further consolidated at higher echelons of command. The soldiers that arrive to support the battalion will not be organic to that battalion, and the commander will have only limited authority over what they do and how they support.

The same can be said for other components of the future battalion task force: signal support, engineer support, chemical support, and so on. Each of these important components of combat power works to support the battalion, but the commander does not have total authority over any of them. In practice, today's battalion commander must employ a mixture of command and anarchic orchestration.

Why does this disunity occur, even at the lowest levels of command? It happens because of organizational initiatives and procedural changes designed to economize. It would be great if each battalion commander could have his own organic maintenance, transportation, supplies, cooks, and medics. But we cannot afford this excess in modern military operations, nor do we need it. Modern technology, built around computer databases and digital communications, has enabled us to economize our logistics so that one consolidated service support outfit can support what used to require many support units.

In other words, control and decision-making authority followed information, in order to stay true to the law of economy.

As we progress toward building and deploying the armed forces of the twenty-first century, we must set aside the myths that have attended command doctrine in the past. The rule of thumb is simple: *Information flow determines decision-making authority.* If we stay flexible and adapt our command doctrine to that equation, we will free ourselves to exploit the other revolutionary aspects of Information Age warfare. We will find, in the end, that we need to balance the competing notions of *command* and *anarchy*.

PART 3
The Laws of War

A first-rate theory predicts; a second-rate theory forbids; and a third-rate theory explains after the event.
—Aleksander Isaakovich Kitaigorodskii

War is a clash of opposing wills, a struggle between beliefs, and victory goes to the party that crushes the enemy's will and destroys his beliefs.
—Introduction to Japanese Principles of War, 1969

12: The Law of Humanity

It is hard to be a coward in the midst of a cavalry charge.
—Jenni Calder

"I was wishing that I came of a more honorable lineage,"
said Caspian.
"You come of the Lord Adam and the Lady Eve," said
Aslan. "And that is both honor enough to erect the head of
the poorest beggar, and shame enough to bow the shoulders of the greatest emperor on earth. Be content."
—C.S. Lewis, "The Chronicles of Narnia"

Law of Humanity: Warfare is an outgrowth of the human soul; all
human conflict is founded upon the nature (physical, psychological, and spiritual) of mankind.

We have seen that the principles of war in their current state do not
suffice as guidelines for modern war. Some require slight revision;
some are outdated; some have never been valid. But if the principles of war are not accurate, relevant, and up-to-date, then what
should we use in their place?

In Part 4, I shall make the point that the most serious problem
with the principles of war is the way they have been used: as aphorisms (truths, rules, prescriptions, and so on). This is not their proper
role. It is as if a director grabbed a stagehand and pushed him in
front of the audience, expecting him to take over the lead role in
the play. Disaster quickly ensues. A good stagehand might make a
poor actor. The same is true of the principles. They cannot serve us
as truths, as we shall see.

But if the principles are not aphorisms, are there any bona fide
truths out there? Are there any statements about warfare that are
eternally true? That have stood the test of time? That can survive revolutions in military art and science?

My belief is that there are three such truths, and I call them laws, to connote their immutability. We are always on dangerous ground when we speak of anything being unchangeable. Of all the so-called enduring truths throughout the history of mankind, most have been ignominiously retired into the dumpster of the ages almost at the instant of their veneration. Man is a very changeable creature, and his recognition of that fact (perhaps even the fear of it) causes him to declare immutability where it does not exist.

But the laws of war *are* immutable. At the very least, I can demonstrate their validity throughout recorded history. And they are unlikely to change, because they are founded on the one aspect of war that remains a constant: the nature of man.

The three laws of war are not equal with one another; there is one preeminent law, from which the other two grow. The first law, and the one that is independent of the others, is humanity: a declaration of the fact that warfare is an outgrowth of the human psyche. From recognition of this simple fact sprouts a correct understanding of conflict in general and warfare in particular.

The other two laws would not exist were it not for the vagaries of flesh and blood. The law of economy springs forth from the essential weakness of man—the fact that man's desires and goals exceed his means. The law of duality—shrouded in mystery throughout the ages—has at its roots the human trait of violent opposition toward other humans.

Together, these three laws give rise to the principles of war (revised). The laws ordain and acknowledge the most fundamental characteristics of warfare and provide a solid foundation of logic and reason upon which valid theory can be built.

The laws are not about techniques, methods, or transitory prescriptions for success in battle. Rather, they describe the true nature of warfare and, in a larger sense, conflict. With a firm grasp on these laws, the student of war can evaluate the countless prescriptions, ideas, theories, and doctrines that he will encounter.

We will begin with the most fundamental truth about warfare: the law of humanity. From there, we will visit the venerable law of economy—that penurious comptroller of Mars. Finally, we will unveil the hidden and mysterious law of duality—a truth that has stalked battlefields from the beginning but which is still a stranger to us.

The law of humanity is the cornerstone of the military art. There can be no other laws or principles apart from it. If you want to judge the validity of any military doctrine, ask yourself first: "How does this doctrine relate to the law of humanity?" If you can find no linkage, you can be sure that the doctrine is in error. Warfare is an expression of the human soul before it is anything else. Guns, tanks, bombs, defense budgets, field manuals, and army chow do not cause war to happen. If we laid all these things neatly on a parade field, nothing would ensue. But throw humans out there—even if you haul off the weapons—and watch the violence start.

It is curious that, although the law of humanity is the foundation upon which all other military considerations rest, it is also the one aspect of war most persistently ignored by military art and science. It is common to hear protests by defense officials and analysts that human factors (especially psychological factors) cannot be accurately studied or simulated. With such claims, we forgive ourselves for paying only lip service to what is, in reality, the centerpiece of the whole business.

But it surprised me to discover that soldiers were lamenting official ignorance in this matter long ago. As Goltz noted in *The Conduct of War*:

> We frequently hear it asserted that psychological considerations should be excluded from the study of the art of war, because mental powers and emotions can neither be gauged nor computed. However, they are of such extreme importance that it would be a great mistake not to mention them. The knowledge of human nature is probably the most difficult, but . . . the most important part of the general knowledge of war . . .

It's time to stop the whining about the problems with studying or simulating human factors in war. Studying war without moral considerations is like studying romance without sexuality. The human heart is the seed from which conflict grows, and regardless of the difficulty, we must learn to deal with moral factors in both the art and science of war.

As of this writing, human factors are not an accepted part of the principles of war within the U.S. armed forces. The absence reveals more about American fighting doctrine than thousands of pages could describe. Americans—soldiers and civilians alike—have chosen to ignore a basic truth of warfare. But why?

The reason is that there are more attractive alternatives to contemplating moral factors. Intellectual laziness craves simplicity and abhors the ineffable complexity of the human heart. If we were to seriously open the books on human psychology as it pertains to war, we would open a Pandora's box of issues to overcome. Some of these revelations would strike at the very heart of some of our deepest-held beliefs about war. In short order, we would have to reorganize major portions of the Department of Defense—an organization currently framed around the fiction of attrition thinking, rather than around truth.

An example—one that will hurt some feelings—is our belief in the efficacy of battlefield fires. No one on the globe believes more fanatically in the power of explosion than the modern American warfighter and his civilian buddies. Twenty years of patient observation has taught me that—official semantics notwithstanding—American military decision makers as a whole believe that the great secret of warfare is to blow up the bad guys at extreme distances. Artillery or air-delivered fires can pretty much solve any military challenge to our nation, if we commit enough funding to indirect-fire weapons, their munitions, and the capability to detect the enemy.

The argument in favor of long-range fires is compelling, well respected, universally agreed to, and flat wrong. Belief in the effectiveness of indirect fires is the single most commonly held delusion in American military history. The firepower that we base our entire warfighting perspective on doesn't work, never did work, and never will work. It is a complete and total farce and a damned expensive one to boot.

Why doesn't it work? Because of the law of humanity. Because we wrestle not against weapons, but against people; we contend not with steel, but with flesh and blood. The destructive force of an explosion will result in a mathematically predictable effect on rolled homogeneous armor plating. But when vectored against people, all bets are off. Some of the targeted souls will be destroyed. Some will flee. Some will capitulate and beg for mercy. Some will stand in irrational defiance, disbelieving their own mortality. Most will recoil in fear and loathing. And eventually, they will think their way around that explosion.

They will hide from detection, shield themselves, attack our weapons, and strive in every possible way to dislocate our fires. They will change the political context, disperse into cities, and dare us to apply our firepower into the midst of noncombatants. They will refuse to be detected, located, tracked, targeted, and assessed. If they die, they will refuse to let us know they are dead, causing us to fire again. They will so complicate the use of fires that in the end, those fires will become overly expensive and utterly useless.

The enemy fears us, and therefore he will adapt.

That is only one side of the moral perspective on fires. There is another, and it is good news for us: The enemy has moral *weakness* as well as strength. Because of this, we have a commonly used but misunderstood military term—*defeat*. What is defeat? It is a condition in which an enemy force has given up the desire to fight. Doctrinaire types will argue endlessly about the technical definition of defeat, but the bottom line is that defeat is 90 percent moral in nature.

Since our armed forces refuse to grapple with moral factors, we are unable to comprehend defeat of the enemy. Consequently, we focus our energies on destruction—the only alternative to defeat. We have removed humanity of warfare out of our calculations, and we are left with a dispassionate, distorted miasma of Lanchester equations that we call military doctrine. As we advance into the twenty-first century, it is time to correct this gross deficiency. The humanity of war can be ignored only at great peril.

The law of humanity requires that we understand and account for both human strengths and human weaknesses. By mastering the factors that lead to strong morale and spiritual resilience, we can develop fighting units that consistently perform beyond their physical capabilities. Conversely, by perceiving the weakness of the human mind and spirit, we can attain an accurate understanding of battlefield defeat: how to avoid it in our own forces and deal it to the enemy. This is an urgent matter, because in any serious future conflict, we will quickly discover how easily enemies can hide from our destructive fires.

The law of humanity is at once the most obvious and disregarded aspect of warfare. Persistent refusal to consider human nature in warfare is itself a peculiar human trait. Dave Grossman, in his book *On Killing*, shows that human beings have an inherent aversion to killing other humans. It is perhaps this distaste for bloodletting that impels us to view warfare abstractly as essentially a political, technical, or even mathematical problem, rather than confronting its true nature.

A failure to recognize the law of humanity requires us to lay aside most of military history, because that history has been ruled by human factors, defying other logic completely. How can we possibly understand the successes of Judas Maccabaeus and his followers in ancient Palestine against the Syrian invaders in the second century B.C.? How can we explain the repeated victories of this outnumbered group of ill-trained, ill-equipped Jewish farmers who faced in battle the world's most powerful army of their day? The law of humanity determined the course of that war, which in turn shaped the development of Christianity and the history of the world thereafter. Attrition equations had no bearing on these outcomes.

Marshal Lautrec came face to face with the humanity of war when his mercenary Swiss captains came to him and demanded that he storm the Imperialists who were bottled up at Bicocca in 1522. Lautrec tried to demonstrate to the Swiss the illogic of their proposition: The enemy were starving and would in all likelihood abandon their strong position in a few days and be at the mercy of the French commander. Instead, the Swiss threatened to march off and leave Lautrec to his fate if he did not immediately attack. His hand forced, the French commander agreed, and the following day, those same Swiss were butchered as they struggled unsuccessfully to grapple with Spanish harquebusiers.

Lee and Jackson reaped the rewards of human weakness as "Fighting Joe" Hooker stood idle at Chancellorsville. Having outmaneu-

vered the Confederates and established himself across the Rappahannock River, he suddenly lost his nerve and waited passively while
Jackson marched unopposed to the uncovered Federal right flank,
there to deliver a devastating blow.

The storm troopers who tore huge gaps in the British lines during the 1918 Spring Offensive should not have paused to plunder
French wine cellars. But they did, and the Allies recovered from their
tactical failures in time to contain the Kaiser's onslaught. Humanity
had overcome doctrine, tactics, and logic.

The German Fifth Panzer Army squeezed through a gap at
Falaise—a gap that should not have been there, but that was provided graciously by the human side of war. Unaccustomed to fighting together, two English-speaking allies—the Americans and the
British—cast away the advantages won at Saint-Lô and allowed the
Germans to fight another day. This failure was not doctrinal; it was
not logical; it was not intended; but it was human, and therefore an
integral part of how warfare unfolds.

The humanity of war will not be ignored. It is time for American
fighting doctrine in all the armed services to look at the problem face
to face.

EMBRACING HUMANITY

What do we do with these embarrassing, irritating, often disastrous
displays of fallen human nature? Conversely, how do we account for
groups of men surrounded, unsupplied, untrained, and ill equipped
somehow defeating trained, professional armies? How shall we pronounce on this routine irrationality that visits battlefields and siege
lines, that collapses well-planned defenses, and trammels the most
calculated attacks?

What do we do with the human side of war?

One solution is to ignore it. It is, after all, the most inscrutable
part of our business. The human heart is slippery, and it evades statistical reduction and scientific analysis like a greased pig. It is rude,
erratic, often ridiculous, and damned hard to quantify. Better, in
some ways, to will it into nonexistence.

But this is not a good answer, despite the fact that most military
doctrines and theory have used it. We may wish the human part of

war aside, but never away. Like something out of an Edgar Allan Poe story, the human heart returns unbidden into our counsels. It will not be ignored. It will ultimately humiliate any doctrine that does not give it its due.

The other solution is hard: build our theory *around* humanity. Rather than giving the human side of war honorable mention, make it central to our perception and articulation of military thought. Ultimately, it should factor into every piece of equipment built, every field manual written, every line of code programmed into our computer simulations.

HUMANITY IN OUR DOCTRINE

If we were to examine official expressions of fighting doctrine in our armed services, we would find some mention—even some exposition—concerning the human side of war. The more technical services—the navy and the air force, specifically—not surprisingly emphasize the emotional and spiritual aspects of war less than the ground services do. After all, it is in their nature that the army and the Marine Corps come literally face to face with human considerations in battle. But despite a grand tradition of sensitivity to moral considerations, not even these components of the American military establishment have advanced far enough.

How do we get the point across in our doctrine? There are reasons to be cautious. In the name of moral and spiritual considerations, some armies of the past have gone too far—imagining that the physics of war count for nothing. The mythical notions of war built around French élan slaughtered hundreds of thousands of French soldiers in World War I. Disciples of the Chinese cult known as the Fists of Righteous Harmony ("Boxers" to their contemptuous Western enemies) dismissed bullets and gunpowder as ineffectual against their mystical powers.

The correct approach is to achieve balance among the factors that cause and condition warfare. Of the three domains recognized by the U.S. Army—physical, intellectual, moral—Western military culture has a firm grasp upon the first two. Our unbalanced doctrines express, study, and practice warfare as largely a physical phenomenon with intellectual factors thrown in. We occasionally admit the moral side of war, but we mostly ignore it and secretly fear it. If, as Napoleon claimed, the moral is to the physical as three is to one, we will find ourselves horribly ill equipped for our profession.

As the armed forces continue to develop information-warfare concepts, we should investigate battlefield psychology as a foundation to tactics. The phenomena of panic, shock, rout, and surrender should become central to the study of combined-arms tactics. Officers should receive instruction in the peculiar sociology of arms, because, as we have seen, defeat in war is most easily understood as a condition of the mind. It is a group phenomenon.

I believe the most direct and effective route to discovering the moral domain of war, and of embracing the law of humanity, is to reform our simulations. In *Art of Maneuver,* I described how simulations used in the Department of Defense (and by civilian agencies in support of the defense establishment) are oriented on attrition equations. Most simulations have little or no moral factors as either inputs or outputs, with the result that our training and requirements determination processes are skewed.

If we were to insist on integrating moral factors into our simulations, several things would result:

First, we would have to commission and resource a defense agency to gather the data and feed it into simulation. This agency would have to become expert in the field of moral factors, and expertise is badly needed.

Second, to study moral factors in war we would have to go to the primary source of information: military history. Of course, many defense officials—both civilian and military—already study military history. Many training courses within the services require study of military history also. The problem is that we are not required to *learn and apply it*—only to read it. If we gave the past its due, we would quickly conclude that the attrition equations upon which our doctrine is based do not pertain to real war.

For these reasons, I believe we should reform our simulations. The effort would be painful, but the payoff would be military doctrines solidly based on the law of humanity, and we could finally overcome the obstacle to understanding and practicing war that has plagued nations since antiquity. The closer we move our doctrine to the human heart, the more effective and valid that doctrine will be, making it easier to avoid the horrible ineptitude of attrition thinking.

Conclusion

The law of humanity includes aspects of psychology, sociology, and even biology. It points us toward considerations of human strength and human weakness. It defines in accurate terms the concepts of defeat and victory. It links easily to the political nature of war. Most important, it reminds us that the physical factors of war are at best secondary.

The law of humanity reminds us of the drama of war. Warfare will expose man's meanest nature and worst characteristics: greed, ambition, indecision, cowardice, cruelty, and hatred. It will also provide a stage upon which man can give expression to his virtues: selflessness, loyalty, courage, restraint, and love.

In the ensuing two chapters, we will see that the law of humanity gives rise immediately to two other laws: economy and duality. These laws have no meaning or real existence apart from humanity. Therefore, at the root of all of our conclusions—of both the laws of war and the principles of war (revised)—lies the law of humanity.

13: The Law of Economy

War is an act of force, and there is no logical limit to the application of that force.

—Clausewitz

If our thoughts are chaotic, so also will our actions be chaotic; consequently discipline of mind must precede discipline of body, and without the cohesion of these two economy of force cannot be effected.

—J. F. C. Fuller

In the realm of conflict and strategy . . . economic principles stand in direct opposition to the demands of conflictual effectiveness . . .

—Edward Luttwak

Law of Economy: The law of economy states that man is weak and lacks resources sufficient to serve his goals in conflict. Further, the supreme danger of armed conflict causes it to be an exceedingly wasteful enterprise. Therefore, to prevail in conflict, one must economize as much as possible.

The most obvious and immediate outgrowth of the law of humanity is economy—an admission and codification of mankind's limited strength and limitless ambitions. We have already seen that the old principle of economy of force retains its importance in twenty-first-century warfare. In Part 2 of this book, I suggested that we elevate economy to a law of war, because it has no logical antithesis. There is an urgent need for economy in warfare and always will be.

The terminology bears mention. I have shortened the older name of this idea to one word, because economy, as a fundamental truth about conflict, is much more than simply the efficient use of *force*. Other nations have chosen to phrase this idea "economy of effort," but this, too, is insufficient. The law of economy reaches far and includes both economy of force and economy of effort. But it also embraces and instructs all levels of war and conflict in many different ways. Therefore, I thought it best to leave the name of the law simply "economy."

UNECONOMICAL FACTORS OF WAR

The law of economy states that, on the one hand, man is resource-poor, but that on the other hand, warfare is wasteful. The law of humanity teaches us the former truth, but why is conflict—especially armed conflict—inherently wasteful? What are the factors that cause waste in war?

First of all, the stakes are very high in armed conflict. Depending on the nature of the conflict, the fate of nations may hang in the balance. More often, however, the strategic issues are not that dramatic, especially since the end of the Cold War. Nevertheless, armed conflict is always dangerous—if not to the nation, then certainly to the individual contestants. Failure on the battlefield will translate to death for some, and lives lost cannot be retrieved. It is this personal danger in war that drives us to win—or at least survive—*regardless of the cost.*

The context of warfare, then, is one in which our compelling need to economize collides with our overriding need to survive. This causes the whole endeavor to be subject to extreme wastefulness. As long as man strives to kill man, the danger of war will lead to lack of economy. But there are other contributing factors that are not necessarily unchangeable.

Ignorance and uncertainty result in wastefulness. In war, there is much that we do not know about ourselves, the enemy, and the environment. We compensate for uncertainty with mass: massed bullets, massed bombs, massed troops, even massed dollars. We cannot afford to be wrong in war; we cannot afford to miss the enemy we are shooting at. In order to overcome our pervasive lack of knowledge, we will hedge our bets with volume. Most of the resources we use in this endeavor will not accomplish their purpose: Most bullets fired miss; most bombs dropped damage nothing important; most troops do not contribute to the defeat of the enemy; most dollars are misspent.

Ignorance breeds waste. But what about truth? Truth, knowledge, understanding, or—to frame it in military terms—information and intelligence act in the opposite direction. Information leads to a *precise* expenditure of resources, and therefore to economy. Indeed, the entire purpose of intelligence in warfare is to economize—to inform our efforts in order to gain effect at the least cost.

We have forgotten the linkage between truth and economy, but it is time to remember it. If twenty-first-century warfare has any theme, it is information. We have vaguely understood that we have nearly reached the physical limits of the technology of violence. We have perfected destruction. In fact, in nuclear weapons, our destructive capability exceeded the needs of warfare, and we had to unbuild it to a degree. As we emerge into the twenty-first century, we can literally kill damn near anything.

Having bumped up against these technological limits, soldiers and scientists began to search elsewhere for effectiveness. Groping for advantage, they discovered information. There is an intuitive obviousness about information warfare: We *know* that it is good to know. The problem is that we have not yet acknowledged to ourselves what we should do with our newly found information: We should economize.

Information leads (or should lead) to *precision*. Precision, in turn, leads (or should lead) to the abandonment of mass. Leaving behind mass, we find (or should find) renewed expressions of economy. In short, information should lead to economy.

The final factor that causes waste, however, is lack of education. Delivering a brand-new M1A2 Abrams battle tank to a group of four men does not create a weapon system. If the men are untrained, the tank is useless. Likewise, giving an army information does not cause effectiveness if the leadership of that army is untrained in how to use it.

What specifically are the educational shortcomings of today's defense establishment? First, we do not, as a rule, perceive the essential linkage between economy and information, as previously noted. Second, we are thoroughly untrained in the dynamics of battlefield adaptation and diminishing effects—phenomena that lead to combined-arms warfare. As a result, we invariably focus on single system approaches, which lead (and must always lead) to attrition warfare—the most atrocious violation of economy that can be imagined.

Every step we take to teach ourselves these things is a step toward the law of economy and a more effective military establishment.

ACHIEVING ECONOMY

Economy results from the mixture of complementary capabilities in such a manner as to exploit battlefield adaptation and diminishing effects. The more precise the mixture, the greater the economy.

In the course of my career, I have read literally hundreds of papers on future weapon systems. Usually, such documents are written by proponents (either official military personnel or civilian contractors) with the intention of championing some system or munition as the key to victory. In the course of these impassioned pleas for funding, the authors typically offer some form of molested military history to buttress their position. For example, one paper I recently read stated that "Since its introduction, artillery has been a significant casualty producer." This assertion, as any first-year history student would know, is utterly false. Artillery existed for hundreds of years before it became a viable killer on the battlefield.

But the biggest problem with such papers is not the misuse or deliberate falsification of history. The worst problem is the "single-system" approach the authors typically take toward future battle. For example, a recent paper praised the fivefold increase in lethality that the army would gain if it fully funded a particular brand of smart antitank munitions. In order to illustrate the supposed lethality, the author conjured up a massed Soviet tank formation (a *what?*) moving toward the inter-German border (toward *what?*) to attack NATO. In a glorious exposition, the writer described the sudden disappearance of the formation while smart antitank rounds, fired from a distant multiple-launch rocket system (MLRS), played havoc with the hapless armored vehicles.

In a similar vein, most proponent writing available today in journals, official memos, or even books, focuses on the lethality of a weapon or munition—typically against a tightly packed battalion of Soviet tanks.

This scenario is a very weak foundation upon which to build such arguments. The most serious issue is not simply the demise of the Soviet Union and its traditional designs on Western Europe. Rather, the salient dilemma is the single-system vision of future conflict, and the focus upon the lethality of weapon systems.

The advanced student of military art learns eventually that the lethality of a given weapon system is a secondary issue. How many tanks a weapon can kill is not as important as another much more critical aspect of the system.

In most human endeavors, and in most professions, success is a stable commodity. The old adage "When something works for me, I stick with it!" perfectly describes the logic of most human pursuits in the arts and sciences. If a farming technique results in better produce, then that technique becomes institutionalized. If engineers develop a more efficient fuel-injection system, then it is adopted as a permanent new feature in cars. If systems administrators develop a more efficient way of managing interoffice communications, it replaces the old way permanently.

But in the professions that deal with conflict, such logic does not work, because the object of the success is belligerent and will react to that success. Like a mutating disease that adapts to antibiotics, the enemy in war will develop defenses against anything we throw at him. Single-system approaches to warfare founder—and will always do so—on this point. The law of economy, conversely, recognizes this inherent adaptive ability of man and builds upon it.

In the technical development of weapons, we achieve economy by focusing not on a weapon's *lethality*, but rather on its complementary effects on other friendly weapons. This is an irony that is fundamentally true and yet completely eludes those immersed in attrition theory. A weapon is relevant to combined-arms warfare only to the degree that its effects can be escaped from.

Weapons of Mass Irrelevance

Since the advent of nuclear weapons, military analysts and more than a few soldiers have become fascinated with so-called weapons of mass destruction (WMD). Theories have grown and prospered. Doctrines have been written. Armies have been reorganized. And some good novels have been written into the bargain. My recommendation is to keep the novels and trash the rest. The WMD fixation is based on fallacy.

Weapons of mass destruction are relevant only when they can find a mass to destroy. Such luscious targets are becoming scarcer than hen's teeth, because mass warfare, despite our love affair with it, is dead. Weapons of mass destruction are, in reality, weapons of mass irrelevance.

Du Picq noted that "weapons are effective only insofar as they influence the morale of the enemy." Now one might argue that a nuclear bomb dropped on Times Square in the middle of the afternoon would certainly influence the morale of New Yorkers. But in a military sense, it would not. As we have noted elsewhere, a military weapon is relevant only to the degree that its effects can be escaped from. *That quality is what makes military weapons politically relevant.* A nuclear weapon leaves no possibility of escape. It is, therefore, not an effective weapon.

Clausewitz instructed us that war is an extension of politics. This essential truth has escaped many of Clausewitz's most devoted students both in the United States and abroad. By focusing on semantics, we have largely missed the great master's point: *Warfare must have a purpose.* Purposeful warfare is a useful and valid expression of man's nature. It is when warfare loses the linkage to a purpose that it becomes an instrument of immorality. This idea of linkage impacts squarely on the design of weapons and military forces.

Warfare is all about imposing our will on someone else. The killing in war is secondary; what we really want is capitulation on our terms.

Experience has taught that it is very difficult to influence the political outlook of a dead man. Therefore, if war is ultimately about politics, then it is ultimately not about killing. We kill some to influence others. If we kill all, then we have removed warfare from the political sphere into a hellish expression of violence and evil. That is not what we are about. Therefore, weapons of mass destruction have very limited utility or effect in war.

At the technical and tactical levels of war, we achieve economy by combining weapons and fighting organizations in a complementary fashion. We focus not on the lethality of each component, but rather on what battlefield reactions that component causes in the enemy. The law of economy calls for us to cause reactions in the enemy's ranks, and then exploit those reactions with other, unlike weapon systems.

At the operational level of war, the same dynamic applies, but the tools are different. Operational warfighting gives rise to joint operations. Just as at the tactical level of war, combined arms is the key to economy and victory, so at the operational level, we must strive for effective integration of air, land, and naval capability. This integration is much easier said than done, because effective joint warfighting requires teamwork, training, and tremendous intellectual energy. Opposing us in these efforts is the natural interservice rivalry that results from our budget process.

At the strategic level of war, a nation gains economy in various ways. Strategy is all about balancing ends, ways, and means. The means of strategy for a nation are diplomatic, economic, military, and informational. According to our current doctrine, these are the elements of national power. This doctrine, however, should be revised to include *intelligence* as an element of power as well, because our intelligence apparatus is a key component of executing our will abroad. In any case, the combination of these elements in a complementary fashion brings about economy.

In practice, however, these separate elements are hard to combine effectively. Our own military doctrine often stands in the way. As we saw in the discussion of objective, one of our national traditions is to speak of using the military only as a last resort. This is an expression of our desire to avoid total war and the horrendous cost associated with it. The "last resort" tradition is viable only as an instrument of diplomatic leverage in a scenario that may lead to total war. When removed from that context, the "last resort" idea loses relevance. To consider using the military only as a last resort means two things: that the other elements of power must operate without effective contribution from the military, and that the military arm will be committed at the point when all other elements have failed. This is an unhappy scenario.

In the twenty-first century, we must not build a foreign policy that uses the military arm as a last resort. Rather, military operations should be considered a full-time component of our strategy and should be combined with the other elements through all stages of a crisis.

It is unrealistic to suppose that American strategy will ever witness a unified, well-conceived integration of the elements of national power. Control of these elements is constitutionally divided. The background and training of military personnel is so utterly different from that of State Department officials that team building is a most difficult art.

We cannot even assert that the president and his administration have total control of all elements of national power. The economic element has a life of its own. The president cannot order people to invest abroad; he has only limited—and mostly negative—control over the economy.

So how do we achieve strategic integration? How do we get to the law of economy with these diverse, disunited tools? The good news is that we do not have to be perfect, only better than our adversaries. And although legal and traditional limits on integration exist, we can overcome them through education, training, and leadership. American strategy in the twenty-first century can and must feature a renewed emphasis on the integration of diplomacy, economy, information, intelligence, and the military forces.

Economy and Simulation

The previous call for a general reform of simulations within the services impacts upon the law of economy as well. The failure to simulate real war effectively has contributed directly to uneconomical practices.

When we began large-scale experimentation in information operations in the army, many of us expected that better visualization of the battlefield would result in a more efficient engagement of the *critical* pieces of the enemy array, and, therefore, a greater economy in terms of ammunition requirements. It seemed intuitively obvious to me that greater accuracy—both in technical and operational terms—would result in fewer required missiles and short tons of ammunition. Imagine my surprise when the experimental units required *five times* the ammunition they used to expend! Suddenly overwhelmed with the vast array of potential targets that they could now see, the experimental units pleaded for more ammunition.

This unprecedented need for large, expensive quantities of ammunition caused logistical nightmares. We simply lacked the trucks, forklifts, personnel, time, and money to feed the commanders the rounds they called for. In desperation, we occasionally had to create artificially high stockage levels to satisfy appetites, and even then the units were not satisfied.

Why did this happen? Why did precision technology not lead to greater economy but instead led to profligacy? The reason is because our simulations failed us. Since neither the software in computer war games nor the much-vaunted OPFOR doctrine recognized the concept of defeat, there *were* no critical targets—at least not in the moral sense. Instead, the experimental units were forced to convert precision intelligence into lengthy target lists. Since defeating the enemy was impossible, destroying him was the only recourse. In the end, the experiments featured apocalyptic destruction of an endless sequence of tightly packed armored OPFOR units. And after four entire armies had been literally annihilated, the fifth army attacked with élan.

We will never live up to the law of economy with such a juvenile understanding of war. Our simulations simply must dispense with the "no-retreat/no-defeat" rule. The entire purpose of fighting is to cause moral failure in the enemy. We kill some to influence others as an expression not only of morality, but also economy. Simulation reform is mandatory to fulfill this goal.

Conclusion

The law of economy is a bedrock of successful conflict. Springing from the nature of man, economy is an immutable truth in warfare. The law of economy has in view a sort of "perfect" warfare: military operations in which available resources are applied with absolute precision, attaining perfect allocative and destructive efficiency with no waste. We shall not attain this perfection, but we can move toward it. Armed with greater information in the future, we *must* seek economy.

14: The Law of Duality

The crux here is making the subjective and the objective
correspond well with each other.

— Mao Tse-tung

A time to rend, and a time to sew;
a time to keep silence, and a time to speak;
A time to love, and a time to hate;
a time of war, and a time of peace.

— Ecclesiastes 3:7,8

There are two aspects to human conflict, but it's damned hard to
name them or explain them. At least, it has been my experience over
the past ten years or so that the dual nature of conflict—while su-
perlatively simple in concept—is next to impossible to teach.

Many have tried, though. Sun-tzu perceived the dual nature of
conflict and decided that it was all about *forces*—one ordinary, one
extraordinary.

Clausewitz understood the dual nature of conflict and described
the *aim* of war (disarming the enemy) as different from the *object* of
war (imposing our will on the enemy).

Hans Delbrück grasped the dual nature of conflict and interpreted it as two complementary strategies: annihilation and ex-
haustion.

When you read about arguments between those who believe in
the use of the military arm to impose law and order or conduct nation building, and those who believe the military should be used only
to destroy enemy armed forces, you are reading about the dual nature of conflict.

When nuclear strategists argued about *counterforce* versus *counter-
value* targets, they were, perhaps unknowingly, discussing the dual
nature of conflict.

When one business fails because it can't compete, and another fails because it has no vision *except* for destroying competitors, it is because neither understood the dual nature of conflict.

When a man can scare off all male competition for the woman he desires but fails to romance her properly and win her love, it is because he does not comprehend the dual nature of conflict.

When a linebacker is trained to knock down offensive linemen through sheer muscle and weight, but then cannot chase and sack the quarterback, he demonstrates failure to master the dual nature of conflict.

In my own writing, I have described the two aspects of conflict in various ways. There is *protective* warfare, and there is *dislocating* warfare. There is *subjective* warfare, and there is *objective* warfare.

It is no overstatement to declare that the dual nature of conflict is the central theme of human interaction. Many have perceived this dichotomy, but a general explanation of this superlative mystery is lacking.

I have come to the conclusion that the dual nature of conflict is so obvious that it cannot be understood without great difficulty. This is a pity, because in my opinion, understanding the dual nature of conflict is the single most important step to succeeding in war, in business, in sports, or even in romance.

So, I begin this chapter with a warning: If you are content with the smell of gunpowder and a few war stories for your understanding of conflict, then don't read this chapter. If, on the other hand, you want to discover the Mystery of the Ages—read on! It will not be easy to comprehend what I am about to tell you. But if you persist, the lights will come on . . . and then you will never read military history in the same way again, because you will have discovered the dual nature of conflict.

SUBJECTIVE AND OBJECTIVE CONFLICT

There are two parts in human conflict: the subjective part and the objective part. I use the term *conflict* in the most general sense of the word: the opposition of human wills. In theory, the dual aspect of conflict applies to human relations, economics, love, and, most important for our purposes, war.

The two phases of conflict, subjective and objective, are fundamental and unchanging. Together, they compose the totality of strategy; when separated from one another—as they most often are, for
they naturally repel each other—they lead to misconception. Although subjective and objective conflict have been detectable
throughout the history of warfare, this fundamental dichotomy is the
single most ignored aspect of war among military theorists and practitioners. Many of the errors in past military thought can be attrib-
uted to the failure to perceive the subjective/objective taxonomy.

Subjective conflict occurs when a competitor fights against his
counterpart, or at any rate, a symmetrical opponent. A tank shooting at another tank, an air force fighting against an enemy air force,
or an army campaigning against another army are all examples of
subjective conflict. One business competing against another, or two
men competing for the same woman are both examples of subjective conflict. Subjective conflict is most simply understood as a con-

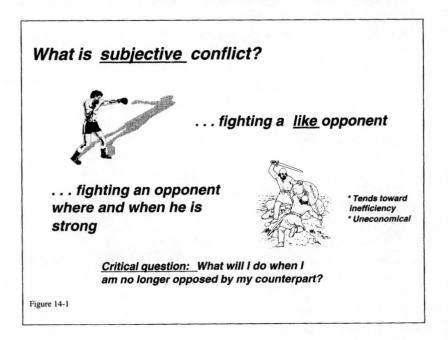

What is _subjective_ conflict?

. . . fighting a _like_ opponent

. . . *fighting an opponent
where and when he is
strong*

* Tends toward
Inefficiency
* Uneconomical

Critical question: What will I do when I
am no longer opposed by my counterpart?

Figure 14-1

test of strength against strength. At the technical/tactical level of war, it consists most often of fighting against a "like" system.

Objective conflict, on the other hand, occurs when an opponent applies combat power against an *unlike* and *vulnerable* aspect of the enemy. A tank overrunning an artillery piece, an air force strafing and bombing enemy shipping, or an army conducting population control are examples of objective conflict. When a business creates wealth and prosperity, or when a man romances a woman, it is an example of objective conflict. Objective conflict pits strength against weakness and vulnerability, or, in a more general sense, it applies energy to something other than competition with a counterpart. At the technical/tactical level of war, it means fighting against an "unlike" system.

This simple and essential dichotomy is relatively easy to grasp, but the implications of it strike to the deepest heart of military theory and practice. Subjective and objective fighting are qualitatively dif-

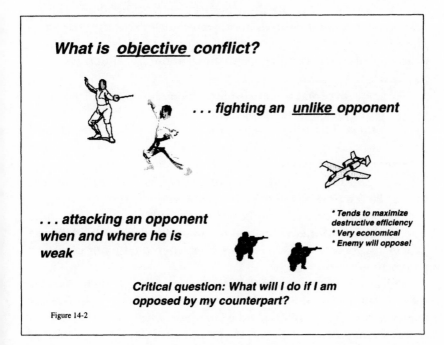

What is _objective_ conflict?

. . . fighting an _unlike_ opponent

. . . **attacking an opponent when and where he is weak**

* Tends to maximize destructive efficiency
* Very economical
* Enemy will oppose!

Critical question: What will I do if I am opposed by my counterpart?

Figure 14-2

ferent concepts. They require from the warrior different skills, different doctrine, different equipment, and different perceptions. Perhaps because of the human passions associated with war, these fraternal twins evoke deep-seated and long-lived mythologies that hamper our efforts to develop our modern warfighting skills today. For although both phases of conflict are essential components of warfare, warriors of the past and present tend to favor one and ignore the other.

The typical soldier (if there is such a person) comprehends subjective conflict. Indeed, he has self-actualized in the practice of it. To the unenlightened, subjective warfare is the *whole* of strategy. If he ever catches a glimmer of objective conflict, he is quick to set it aside as a nonmilitary—even distasteful—phenomenon. This soldier, if we were to analyze him, would reveal an emotional, almost spiritual adoration for subjective conflict and a revulsion for the "other stuff." If he becomes a leader, he will train his subordinates and prepare them for subjective conflict. He will plead with his betters for more resources to increase the army's subjective capabilities. He will relentlessly define the threat in subjective terms and cultivate a singular fascination for the mirror image of himself. If ever asked to prosecute objective conflict, he will bitch about it . . . but try to muddle through.

The subjective infantryman orients on enemy infantry. He is most proud of his rifle marksmanship, his hand-to-hand combat skills, and his toughness. The subjective pilot longs to fly fighter aircraft and shoot down enemy fighters. To him, a war is virtually over when the skies are clear of enemy jet fighters. The subjective general organizes his joint task force with the goal of destroying the enemy armed forces. He emphasizes destructive firepower and the garnering of overwhelming combat resources. He gives little thought to population control or to the imposition of law and order on the defeated provinces. Sometimes he is embarrassed when the finality of victory eludes him.

Typically, he grumbles, points to the burning enemy tanks, and then writes doctrinal manuals that proclaim that "the ultimate objective in all military operations is the destruction of enemy armed forces." The subjective warrior ever fails to ask a critical question of himself: *What will I do when I am no longer opposed by my counterpart?*

The objective warrior is no more successful. Often, this type of perspective is found in civilian officials, but there have been many military men who have embraced the objective viewpoint. Warfare, to the objective thinker, is a matter of applying combat power against weaknesses and vulnerabilities, or, in a strategic sense, against other—nonmilitary—entities. Indeed, the objective viewpoint may not even consider his use of the military as conflict at all: He may opt to use the military arm to distribute disaster relief supplies to devastated areas, or to protect other organizations that are doing so. He may look to the military to monitor and conduct stability operations, or conduct population control. In short, he views the main task of the armed forces as operating in support of a political, economic, or cultural goal.

Is objective conflict really conflict? Can we really consider a business creating wealth or a man romancing a woman to be conflict at all? The answer is yes, because the objective strategy pits our abilities against the status quo. Our business creates wealth where there was none before; our lover creates passion where before there was nothing. Objective conflict is a clash between the possible and the actual.

Objective thinking also manifests itself at every level of conflict down to the technical/tactical level. To the objective soldier, asymmetrical application of combat power is the *whole* of strategy. In his visionary zeal and his well-intentioned genius, the objective thinker often forgets that the enemy "has a vote" and that his efforts might be opposed by a symmetrical opponent. The objective infantryman considers the killing of tanks, or the hunting of guerrillas, or the capture of an international terrorist to be his main business. He does not often think about or train to take on similarly equipped infantrymen. The objective pilot—perhaps a rogue among his associates—is a specialist in bombing ground or naval targets. He is an expert in the complicated processes of targeting tanks, supply dumps, or shipping facilities. Armed forces that are oriented on objective conflict organize themselves around the avoidance of armed conflict and instead seek to use their rifles and guns to influence enemy policy, economy, and culture. If suddenly faced with a mirror image of himself, the objective warrior is surprised and subjected to disastrous defeat. Uncomprehending, he casts about for someone to blame:

most often the intelligence apparatus. The objective warrior, like his subjective partner, has failed to ask himself a critical question: *What will I do if I am opposed by my counterpart?*

From a thorough understanding of subjective and objective phasing, we can harvest keen insights on other principles of war. If we grasp the dual nature of war, we are not deceived into thinking that mass applies to *warfare* as a whole, but only to the subjective part (if then). We must "unmass" to conduct objective conflict. Likewise, the subjective/objective duality reveals the deeper aspects of offensive and defensive operations. The attack is, by nature, objective. The defense is fundamentally subjective. Finally, the subjective/objective taxonomy shows us the balance between the need for dislocation (an expression of objective warfare) and confrontation (subjective warfare).

This dichotomy—this bloody and deeply ingrained dual failure to account for *both* phases of conflict—pervades military history and theory. The force of this two-way intellectual bias confounds the neutral observer. It is no overstatement to note that there are subjective thinkers who would sooner sacrifice their lives than admit to the bona fide existence of objective warfare. Likewise, there are officials who actually lack the intellectual capacity to comprehend the need for subjective capability in armed forces.

A clear manifestation of imbalance is the armed forces' approach to information operations. Joint doctrine writers have defined information warfare as "actions taken to achieve information superiority by affecting adversary information, information-based processes, information systems, and computer-based networks while defending one's own information, information-based processes, information systems, and computer-based networks." Verbosity aside, this definition is completely inadequate. It conceives of one army employing information assets to attack and defend against the enemy's information assets. This vision is totally symmetrical, totally subjective, and totally useless.

Information warfare—if it is to be meaningful at all—must ultimately impact upon the other aspects of warfare. Information warfare is not just about destroying enemy information; it is about making the friendly force move faster, shoot better, and protect itself more economically. It is about slowing the enemy, disrupting his operations, demoralizing him. It is about political penetration of a theater of war. It is not simply digital shadowboxing.

This definition came about because of a pervasive lack of understanding concerning the dual nature of war within the armed services. Effective thinking requires a strong balance between beating the enemy subjectively *and* objectively.

It is from this holistic perspective on the military art that we must utterly reject the U.S. Army's official insistence on the destruction of enemy armed forces as the ultimate objective in war. This assertion is a time-honored, well-respected load of hogwash. The ultimate objective of military operations is the application of combat power to enforce a policy of some kind, whether cultural, political, economic, or related to security. The destruction of our enemy counterpart is a necessary, vital component of the whole of strategy, but it is at most half of the equation.

Alexander the Great was a master of both aspects of warfare. One of the key reasons he conquered Persia is because he had an enlightened policy for the lands he won. To him, war was not just about beating the Persians. It was about *building* as well as *destroying; ruling* as well as *defeating*.

STRATEGIC ERROR

Failure to comprehend the essential duality of conflict leads not only to errors in strategy, but to *catastrophic* errors. As the debate over the future of American military strategy continues, this failure is manifest in what we have termed the "Precision Strike" school of thought.

As we have seen, the precision strike argument is that with modern missile technology, fused with sensor and command and control technologies, future war will be characterized by long-range precision destruction of national enemies without significant direct-fire combat. Visionaries of this school foresee a future crisis in a hot spot abroad. On the heels of the emergency, land-based and sea-based aircraft, in combination with smart missiles, quickly destroy enemy formations and installations with impunity. The need for bloody direct-fire combat is thus avoided.

This type of scenario is nonsense, and it reveals a frightening lack of understanding among those who propound it. In order to believe in this concept of war, we must concoct the most predictable and inane threat force imaginable. We must insist that all noncombatants leave the vicinity. We must remove that densely packed mass of armored vehicles from any nearby cities, jungles, mountains, or any other terrain features onto a flat plain. A desert will do nicely, thank you. As long as we can convince the enemy to fight us in this manner, the precision-strike school will prevail.

Of course, anyone vaguely familiar with emerging technology could attack the precision-strike scenario on the grounds of feasibility and cost also, but there is a more fundamental problem with this errant vision.

This school of thought does not grasp the dual nature of conflict. It is focused strictly on the subjective aspects of war, to the utter exclusion of the objective side. The power to destroy something is not equivalent to the power to control people. Real war and effective strategy must combine these two concepts. Success must be defined both in subjective and objective terms.

Goltz perceived this truth long before the invention of long-range missiles.

> Theoretically, it is quite conceivable that a state may destroy the organized military power of another nation and overrun a great part of its territory, and yet not be able to bear for long the sacrifices [required] to grant a comparatively favourable peace to the defeated state. This is frequently lost sight of, and the destruction of the enemy's main army is taken as being synonymous with the complete attainment of the object of the war.
>
> It is only when the two opposing states are somewhat similar in national characteristics that the ultimate object, namely, the enforcement of the desired peace, can at once be attained by the defeat of the enemy's main army.
>
> When, on the other hand, the inner nature of the belligerent states is different, the defeat of the enemy's fighting forces and the enforcement of peace will but seldom coincide.

It is dangerous and ineffective to assert, as many in the army have done, that we should focus our efforts on "the hard fight"—that is, the subjective engagement of the enemy. It is often said that if we can prevail in combat, then peace operations are easy. People who believe this simply haven't worked the problem of stability operations. Objective warfare is radically different from its subjective counterpart, and oversimplifying one or the other is the path to defeat. As Lt. Col. Kenneth Pritchard noted in a recent article:

> Successful civil-military operations in the 21st century will require new skills in relationship development, negotiation, mediation, arbitration, world political awareness and a big-picture, long-term perspective. Also imperative are skills in civil information operations (even military events in remote areas take place under the watchful eyes of CNN) . . .

The law of duality is an essential part of striving for economy in war. When a strategy for defeating the enemy is couched only in subjective terms, success will almost certainly elude the unfortunate army charged with its prosecution. Commanders who fail to perceive the other half of warfare will have no recourse but to try harder through subjective approaches. The American army in Vietnam is a classic example. What we could not achieve through an objective application of combat power, we tried to snatch through relentless "smash-mouth" warfare with the Vietcong. Subjectivity eventually grew into a horrifying lack of economy, and the nation simply ran out of money, time, and patience. The greatest expression of the law of economy is the law of duality.

IMPLICATIONS OF THE DUALITY OF CONFLICT
Once we understand the dual nature of conflict, the next step is to apply that knowledge. There is no aspect of military art and science that can advance properly without first being rooted in this concept. I shall touch briefly on only two of those areas: requirements determination and leader development.

Requirements Determination
The military services are charged by law to develop requirements for the future. Those requirements eventually translate into materiel-acquisition programs, with associated funding, doctrine, organiza-

tion, training, and leader development implications. To the layman, the bottom line is that the armed services determine what weapons and equipment they should invest in, and Congress ultimately authorizes procurement. The processes whereby these decisions are made are complex and do not make for interesting reading, so I will not discuss them. Instead, let's cut to the heart of the matter: How should the dual nature of conflict influence the development of weapons?

Let's start with the conclusion: *A weapon that does not have dual capability is a bad investment.*

On the battlefield, at sea, or in the air, a weapon system will have two types of targets: targets that resemble it and targets that do not resemble it. Or to put it another way, each friendly weapon will face both *like* systems and *unlike* systems. Good weapons prevail against both types. Ineffective weapons prevail against only like or unlike systems, but not both.

Because subjective thinking prevails among military professionals, weapons tend to be developed with like-system thinking in mind. Tank designers try to optimize the tank to fight enemy tanks. They structure their developmental programs around estimates (and, by extension, *over* estimates) of enemy tank capability. They analyze these estimated enemy capabilities in the context of battlefield scenarios that lead to symmetrical engagements of tanks against tanks. Although it is frequently unnoticed, this type of requirements determination leads to spiraling costs as the dynamics of attacking munitions and defensive measures compete with each other. Ultimately, the tank can become outrageously expensive, too heavy for most tactical environments, and virtually useless in any except purely symmetrical engagements.

When left to this design methodology, the tank, as an example, becomes encased in a confining body cast of doctrine, tactics, and training focused ever more narrowly toward the mirror image of itself. With each step in this development, this shadowboxing weapon system loses utility against other, unlike systems. The finite number of main gun rounds becomes composed more and more of tank-killing munitions, and less and less of munitions useful against infantry, buildings, or other targets. In anticipation of incoming tank rounds, the tank's armor gets thicker, heavier, and/or more expensive. The weight goes up, and the fuel economy goes down. In the end, we are left with a barely mobile, muscle-bound hulk capable only of destroying an exact replica of itself . . . but just barely. Such a system—frustrated in the end with the illogic of its own symmetry—has no recourse but mass warfare to overcome its self-made contradictions.

According to the subjective logic illustrated earlier, this tank has forgotten to ask itself the question: *What will I do when I am no longer opposed by my counterpart?* Inability to answer that question makes the tank an ineffective weapon and a bad investment.

Why is this "subjective" tank a bad investment? Because if the enemy chooses to continue opposing our developments symmetrically (i.e., by engaging us in a cycle of competitive tank improvements), then expense will skyrocket. If, on the other hand, the enemy chooses to forego symmetrical developments, then he will ipso facto remove from the battlefield viable targets for our tank. Since the "subjective" tank lacks capability against other, unlike systems, it has no useful role.

The same can be said for ships and aircraft, or for any killing system. If it is designed around a mirror image of itself, it will be too expensive and ultimately useless on the battlefield.

What is the solution? To design weapon systems simultaneously against both like and unlike systems. This should not be a difficult thing to attain. After all, *all weapons are originally designed with unlike targets in mind.* The tank, when first developed in World War I, was aimed at busting trenchlines and killing infantrymen. It was only after the developmental cycle advanced into World War II that serious thought was given to the antitank role of tanks. Likewise, aircraft entered the realm of military art first as a tool for observing the battlefield, later for bombing and strafing ground targets, and only years later as a means for shooting down other aircraft. Ships were first built to transport goods and soldiers. Later, they were equipped with rams for opposing other ships.

The problem is that new, innovative weapons eventually become institutionalized for further development and production. And when they do, they typically get sidetracked by the fear of competitive development by the enemy. This leads to an overemphasis upon symmetrical, subjective capability and a withering away of asymmetrical, objective utility.

Leader Development and Training
The law of duality must be cultivated within our leaders before it will bear fruit. As we have seen, an understanding of the dual nature of conflict does not come naturally or easily. It must be taught repeatedly. This law is always at risk in the minds of soldiers, businessmen, sportsman, and lovers. It is normal for people to emphasize one aspect of conflict over the other. Left to themselves, most will eventually lose sight of the dual nature entirely.

Military leaders generally perceive the subjective aspects of the problem. Current army training and indoctrination certainly points officers and sergeants toward the elimination of symmetrical opponents. The weakness in current leader development is in the training of objective conflict. But how do we get officers to think objectively?

The key is to widen their experiences and reading. Objective understanding of warfare comes through the liberal arts—history, philosophy, literature, and so on. It is by learning about things *nonmilitary* that the officer perceives the relevance of military things. If Douglas Haig is typical of senior military leaders who lack the imagination to avoid bloody attrition fighting, part of the problem is revealed by the field marshal's biographer, Brig. Gen. John Charteris, as he commented on Haig's reading life:

> [Haig] had strangely little learning; his military work absorbed him, and he only glanced at other subjects, never studied them . . . he read few books and never a single novel.

> Kitchener had wide interests outside military matters—art, science, politics, literature claimed their share of attention; Haig's whole mind and life were concentrated on the army and on war.

An objective understanding of war depends upon contemplation of matters other than the sword. It has been posited that creativity is essentially the ability to mentally connect two ideas that were previously unconnected. If so, then the potential for creative military thought lies in the officer who cultivates an appreciation for nonmilitary things. The ability to imagine an innovative solution to problems in war rests not only upon military competence, but also upon a wide interest in other areas of life.

It is perhaps this benefit of nonmilitary pursuits that prompted the requirement for officers in the armies of the eighteenth and nineteenth centuries to be "men of liberal education"—a phrase often found in treatises of that period. Today, exposure to literature, philosophy, history, and other liberal arts can be a much needed balance against the technical education of our officer corps. Creative thought can never come from the study of one subject; it is the bringing together of unlike ideas.

Conclusion

All conflict has a dual nature, and no person, organization, or nation can be successful in conflict without accounting for both aspects.

The successful contender is a subjective and objective fighter. He contemplates simultaneously the like and the unlike. He thinks about the problem of being opposed and the equally difficult problem of *not* being opposed. When ensconced in a subjective fight, he searches for objectivity. When enjoying an objective advantage, he watches for indicators of a subjective opposition.

The warrior who knows about the dual nature of conflict is double-minded and victorious.

PART 4

The New Principles of War

Time and again, where a radical change in equipment, doctrine or force structure is concerned, one finds a gestation period of between 30 and 50 years or more between the technique becoming feasible, or the need for change apparent, and full-scale adoption of the innovation.

—Richard Simpkin

15: The Arguments of War

As a vision of strategy emerged out of the shadows of words read, problems investigated, and events experienced, I found that its content was not the prosaic stuff of platitudes, but instead paradox, irony, and contradiction.

—Edward Luttwak

War is the interaction of opposites.

—Clausewitz

Nothing has such power to broaden the mind as the ability to investigate systematically and truly all that comes under thy observation in life.

—Marcus Aurelius

As a father of five, I have built a lot of cribs, bicycles, and toys. Unfortunately, I have yet to mature to the point where I read the instructions first. I try to do it right, but when I empty the box of parts and hardware onto the floor, I get overwhelmed with the desire to get started without reading the novel-sized instructions, complete with warnings in seven different languages. Besides, it is, quite frankly, an insult to my manhood to suggest that I need anyone telling me how to do something.

Consequently, we have seen some odd-looking cribs over the years. Some of the drop sides tended to malfunction at inopportune moments: One of them locked up tight and never moved again; another resembled a fully functional guillotine rather than a sleeping place; and one of the cribs simply collapsed into a neat stack onto the floor moments after I snatched the wriggling infant into my arms. I never did figure that one out. Finally, with the fifth child, we opted for a mattress on the floor.

Most men don't like instructions—not for crib building and not for war. But in the matter of the principles of war, instruction is required. One of the reasons we have had to revise the current principles is because they have been misused for many years, and the misuse has warped them. Before we can learn anything from them, we must first understand what to do with them. Used correctly, they lead to victory in war.

First, the warning label: *Do not use these principles as aphorisms!*

An aphorism—whether disguised semantically as a principle, prescription, rule, or guideline—is simply a *truth* of some sort. It is a statement of what is or what ought to be. I have already admitted to the existence of a very small number of bona fide laws (a highfalutin word for an aphorism), but beyond the three mentioned in this book, I doubt if there are any others pertaining to warfare. Above all, we must not consider the principles to be aphorisms, because *they cannot serve us in that role.*

To determine the correct way to use the principles of war, let us hear from two radically opposed schools of thought concerning them. I call them the Aphorism School and the No Principles School. The former is the traditional body of thinking that says that the principles of war are useful guidelines to action in military operations. This is the official position of the U.S. armed forces. I know of no one who is so narrow-minded as to call for utter and slavish obedience to the principles. Rather, most advocates of the Aphorism School admit that some flexibility in application is required.

The No Principles School of thought says that there are no valid principles in warfare, because every situation is unique. Each war, campaign, and battle is governed by innumerable, complex factors (military, political, environmental, technological, and so on), with the result that nothing learned in one conflict can apply reliably in another. The key to good generalship according to this perspective is to quickly perceive the critical factors of the operation and use them to advantage. This school of thought typically heaps scorn on the principles of war and their adherents.

Both of these opinions are reasonable, and advocates of both sides of this issue can point to historical examples to prove their contentions. In fact, we discover the truth not in one school or the other, but in between.

The Aphorism School is in error, because, as I hope I have demonstrated in Part 2 of this book, the principles of war have, in practice, been subject to numerous exceptions and qualifications. Above all, they trade off against one another. Telling a future general to stay true to both security and surprise is like telling him to face simultaneously both right and left. Insisting upon clever maneuver while requiring simplicity is like demanding a vegetarian roast beef sandwich. For this reason, we will cultivate confusion and, ultimately, a frustration with the principles if we regard them as rules or even as guidelines—for they are not.

The No Principles School is on even shakier ground. To proclaim belief in the utter uniqueness of each military situation is completely absurd. It is tantamount to saying that we can learn nothing from history, or even from personal experience. We *can* learn and apply that knowledge to future conflict, because situations are only *somewhat* unique. Human conflict is governed by laws, and illuminated by principles. Although it is important to recognize uniqueness in each situation, we must also perceive similarity.

The correct way to use the principles of war is to treat them as *arguments*. They are not valid as prescriptions; they do not express truth at all. Instead, they describe *categories of thinking*. They instruct the soldier what to think about, but not what to conclude. Future conflicts will be similar enough to past conflicts as to allow application of laws and consideration of principles; but those conflicts will also be unpredictable enough to require a unique synthesis on the part of the commanders who face them.

What do we want from our principles of war? We want them to provide a solid, reliable, and comprehensive instruction to our warriors concerning the categories that they must think through as they plan and fight. We want them to be dynamic tools for the trenches—useful mentors even in the midst of a life-and-death struggle. But how do we get there?

We learn to argue. Not with each other, but within ourselves. Ingenuity, creativity, and innovation on the battlefield are born of argument. The commander who acquires the intellectual discipline to argue with himself can break free from staid, ineffective thinking and routinely develop creative and productive battlefield solutions.

The new principles of war are intended to frame argument. As we will see in the next chapter, the principles are expressed not as one-word aphorisms, but rather as two-word arguments. They describe a given category about which the soldier must think, and they indicate the opposite sides of the argument.

Those familiar with philosophy will recognize this form of logic as *dialectic*. Dialectic logic is all about arguing. In fact, the word *dialectic* is a cognate of "dialogue." The successful soldier of the future must be skilled at talking to himself. He must learn how to use dialectic logic.

A dialectic begins with a *thesis*. A thesis is simply a statement or proposition. For example, I might propose to my wife that "The family absolutely needs to have a brand-new Corvette." I use this example, because I have found that the best way to effectively argue is to start with an extreme position. The more extreme the thesis, the more abrupt and dramatic will be the reaction.

My wife recoils in shock and responds with the next step in dialectic logic—the *antithesis*. "We do not need a new car at all. The old one is fine." As you can see, the antithesis is not merely a disagreement, it is a *total* disagreement. We are not arguing about the color of the Corvette, nor even about whether to buy an American or foreign car. Instead, our positions are extreme and utterly opposite.

The final step in dialectic logic is the *synthesis*, or the bringing together of opposites. Rather than my insisting that "It's my money and I'll do what I please!" or my wife shouting "If you ever mention this again, I'm leaving you!" the synthesis tries to find a middle ground. We do something more than sticking with the old car, but less than diving into debt with a brand-new, expensive sports car.

This process is what the principles of war are meant to do. They teach us what to argue about, and they define the end points of those arguments.

Dialectic logic is, in my experience, a superior way to think on the battlefield. It leads to far better conclusions than intuition or undisciplined thinking. Let's illustrate why with another example, this time from the business world.

Let us suppose that we are all on the board of directors of an investment company. At issue is our portfolio of mineral stocks. The question before the board is what, if anything, we should do with the stocks. How do we go about exploring this question and reaching an effective solution?

One method would be to chat about the issue in an undisciplined discussion.

"I don't know," one director begins. "I think maybe we should shift some of our assets toward short-term, high-risk investments."

"Well, I think we should watch the market for a few days," says another.

This type of discussion lacks rigor and aims at an early consensus. Typically, intuition and undisciplined thinking lead to mediocre solutions. The problem is that the board has not defined the end points of the argument, and as a result, the options they are contemplating are too limited.

The dialectic approach would be to appoint one of the directors to take up an extreme thesis: "We must immediately divest ourselves of all mineral stocks, because they are about to plummet in value."

Another director is assigned the antithetical position: "The market fluctuation is going to be temporary, and we should buy up as many mineral shares as we can afford, even selling off other assets to generate the revenue."

Now we have a rigorous argument. We have two extreme positions, and we have, therefore, a wide range of middle ground options to choose from. To strengthen the process, the first director should assemble all the evidence he can find to support his position that we should sell. Likewise, the second director should research his contention that we should buy. As the rest of us listen to the give-and-take between these two, we can critically analyze the evidence given. We will find that the second director's facts are questionable, perhaps, and so move more toward selling. But even if we tend to agree with the first director, we will find that his extreme position begins to migrate beyond common sense, and that rather than selling all mineral assets, we should probably focus on the ones with the highest risk in the short term.

We will develop a synthesis—a solution in the middle ground that acknowledges and accounts for the unique factors of this particular problem. We didn't begin the process with consensus; in fact, we deliberately destroyed premature agreement by constructing a dialectic. The consensus that we finally reach will be founded upon solid evidence, not upon office politics and the natural fear of confrontation.

In a similar manner, the soldier on tomorrow's battlefield will be confronted with difficult, complicated problems to solve. He may get shot at while he's trying to think, too. Because the decisions the leaders make will ordain the fates of so many—on both sides—we must ensure that those leaders are equipped with the very best doctrine and training. In our mind's eye, we must consider the specter of Douglas Haig peering across no-man's-land, regretting the deaths of the thousands he sent to destruction, and then concluding that next time we must charge more vigorously. We want no more such scenarios. Instead, we want our leaders to think dialectically.

How do we create genius? We relentlessly train our officers and soldiers to argue with themselves. We teach them to oppose their own thinking, rather than letting the enemy do it for them. They must construct healthy arguments, think through the extremes, and select a creative, logical option. The alternative is to argue with the enemy using bullets and blood.

I cannot overemphasize how powerful dialectic logic can be on the battlefield. Nor can I express how conspicuously absent such thinking is in the military. It is unfortunate and certainly unintended, but when a general walks into a room, consensus follows like one of those annoying little lap dogs. Heads bob up and down in vacuous agreement, and dialectics flee.

Military hierarchy has a logical foundation in battle. In the extremely time-competitive, life-and-death struggle between two enemies shooting at each other, we have no time for argument—either with each other or within ourselves. Someone is shooting at us; something must be done. The captain yells, "Follow me!" and we follow. He is in charge, because we know that rapid decision making and faithful obedience prevail in gunfights.

But the further we remove ourselves from the sound of those guns, the less the military hierarchy makes sense. Today, in the U.S. armed forces, decisions are made by the hierarchy, sometimes in the most arbitrary and uninformed way imaginable. What was originally a battlefield expedient in the ancient world has become a bloated, inefficient mechanism for the squashing of independent thought.

The purpose of this book is not to call for the disassembly of the military chain of command, but rather to urge a new form of leader development: dialectic logic. But to cultivate the art of argument within our leaders, we must distance ourselves from the obsequiousness that follows rank in the armed services. Only the leaders themselves can make this happen. The greatest legacy that a leader can leave behind is a subordinate who is not afraid to think for himself.

Could tomorrow's leader step into Doug Haig's shoes and come up with effective solutions? Could he argue within himself and use the principles of war to graduate beyond the "charge harder next time" strategy? What might he be able to create that Haig could not create?

The logic might proceed like this: I am expending soldiers at a highly uneconomical rate. Something must be wrong with my approach to defeating the Germans. The law of economy demands another solution. The law of duality instructs me that I am fighting subjectively, with no objective payoff.

I have been fighting according to classical methods of mass warfare. I must argue with myself against all these assumptions. Let us assume for the moment, that mass warfare is wrong. Is it possible to dislocate the enemy trenches? Machine guns? Artillery?

Can we achieve a positional dislocation? Perhaps if we attacked from the rear . . . but how do we get there?

Can we functionally dislocate? We would have to apply a new technology . . . the tank, perhaps?

Can we cause a moral collapse? Probably, but only after inflicting some measure of physical defeat and destruction.

What about changing the time dynamics? Can I attack the enemy when he is unready? What causes him to prepare? Ah! Our prolonged artillery bombardments! And his detection of large troop movements. What if we forego long preparatory fires and charge across no-man's-land with a handful of troops at night?

Let us imagine, for a moment, great success with these tactical measures. Let's assume that we crack the enemy's front lines. Then what? How can I turn a subjective, tactical advantage into an objective, strategic payoff? I must develop a highly mobile capability to exploit temporary breakthroughs. Perhaps a force built around the internal-combustion engine . . .

I must organize the battlefield into two categories: what I know about and what I don't know about. Those things that I know about, I should convert into effective, authoritative decisions and directions to my subordinates. But those aspects that I do not and cannot know about, I must defer to my subordinates to decide. I must organize my command philosophy around the distribution of battlefield truth.

This has been an unfair example. I am concocting solutions based upon the history that I know. All these solutions to the deadlock on the western front were in fact developed, but not by Haig. Some came from the Germans, others from the Allies. They were not developed through pure intellect and theory, but through bloody experience and failure.

Could Haig or his contemporaries have done better? I must conclude that they could have, or I must abandon my profession. The military art is an intellectual art. The laws and principles of war can obviate disproportionate effusion of blood if we are determined to do so. To that end, I offer the new principles of war for the Information Age.

16: The New Principles of War for the Information Age

The military system ought to rest on good principles that experience has shown to be valid.

—Frederick the Great

No laws or principles which do not accurately describe processes and relationships can be relied on to predict or control them. To control, we must predict; to predict, we must describe accurately; to describe accurately, we must omit no essential element of the processes concerned.

—Reginald Bretnor, Decisive Warfare

Human conflict is governed by three laws, as explained in Part 3: the law of humanity; the law of economy; and the law of the duality of conflict. Of the three, the law of humanity is the foundation: it is the independent law, upon which the other two are built. Economy and duality emanate from humanity. These three laws, when applied to conflict in the Information Age, give rise to seven principles of war, one of which is an independent principle, and six of which are dependent upon the first for application.

THE PRINCIPLE OF KNOWLEDGE AND IGNORANCE
"Knowledge" concerns the information that we have or intend to have about ourselves, the enemy, and the environment. Ignorance is the converse of knowledge, and it deals with what we do not know, what we cannot know, or what we choose not to know. Both knowledge and ignorance have dominated warfare throughout history, but Information Age warfare has adjusted the balance toward knowledge. Commanders and soldiers in the Information Age will be relatively more aware of their surroundings than were their predeces-

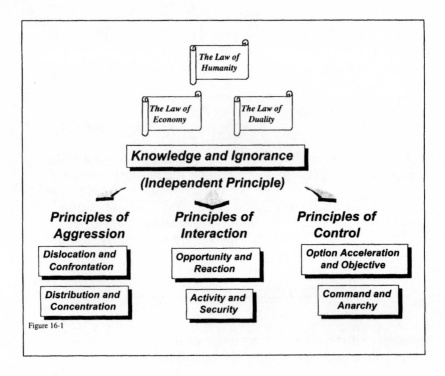

Figure 16-1

sors in other ages. Because of this, application of the other principles will change significantly.

Nevertheless, one of the skills of future warfare will be to actively manage what we know and what we don't know. The principle of knowledge and ignorance is based on the proposition that, given time, a commander can know everything about a given conflict. However, because conflict is time competitive, he must choose not to know certain aspects, and he must adapt his activity to most efficiently manage that ignorance.

Therefore, warfare governed by the principle of knowledge and ignorance does not suppose that knowledge is good and ignorance is to be avoided. Rather, we understand that information has a cost associated with it. Truth on the battlefield costs time, lives, and supplies. Ignorance is free. The acme of skill in the Information Age is to manage what we know and what we don't know, and to balance our knowledge with activity. In this respect, the principle of knowledge and ignorance is conditioned by the law of economy.

The law of economy seeks to minimize cost. We have seen that "cost" in the military art is figured in various currencies: political will, money, supplies, time, and lives. As we stand on the edge of the twenty-first century—with all the technological implications of the Information Age—we have a fundamental choice to make regarding how to achieve economy in future warfare. All armies, when faced with information technology, must likewise decide between two courses toward economy.

We could, on the one hand, choose ignorance. Ignorance is cheap and in ready supply. I am not being sarcastic in declaring this to be a potential choice for us. There are good reasons—most having to do with technical feasibility and costing—for us to reject (on a relative scale) the widespread development and fielding of information technologies throughout the armed forces. We would, in very short order, save millions of dollars and retrieve badly needed training time.

The problem is, as we have seen in Part 2, that in the context of ignorance, all the other aspects of war are very expensive. If we save money on information technology, we will pay tenfold (at least) in other areas. Ignorant armies must ultimately resort to mass warfare. They must more often *confront* enemy strength in attrition contests rather than *dislocate* it. They must constantly *react* to the enemy, rather than capitalizing on *opportunity*. They must spend enormous resources in *securing* themselves against the unknown, reserving little for positive *activity*. They must unify their actions through the early and inflexible selection of an *objective*, rather than enjoy the more effective process of *option acceleration*. In short, ignorance is penny wise and pound foolish.

The other course is to choose for knowledge. Information technologies are among the most expensive in the military realm, because they exist only as a *system of systems*. Information dominance in future warfare is *not* built only upon the computer. It does not rely solely on communications. It is not based only on sensor technology. It is not merely a training innovation. It is *all* of these things— and much more—welded together by doctrine. In the short term, knowledge is a pearl of great price.

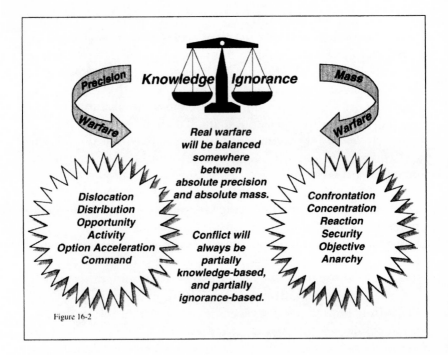

Knowledge **Ignorance**

Precision **Warfare**

Mass **Warfare**

*Real warfare
will be balanced
somewhere
between
absolute precision
and absolute mass.*

Dislocation
Distribution
Opportunity
Activity
Option Acceleration
Command

*Conflict will
always be
partially
knowledge-based,
and partially
ignorance-based.*

Confrontation
Concentration
Reaction
Security
Objective
Anarchy

Figure 16-2

But the good news is that short-term investment in knowledge yields extreme economy in every other aspect of warfighting. The scripture declares that "Truth shall make you free." As it happens, this spiritual dictum is true in the military sense as well. Truth frees us from mass warfare, from bloody struggles over initiative, and from the tyranny of securing against the unknown. The sword that is guided by truth is precise, accurate, relevant, and swift. And, at the end of the struggle, it does not beggar the nation that wields it.

My own belief is that information technology is a good investment, but the many decisions that will build or emasculate knowledge-based warfare in the future force are political. As I stated in the preface, it is not my intention to wade into defense politics, but rather to explain the theory and practice of war as I have learned it. The point I desire to leave the reader with is not what political decisions should be made, but rather what the implications of those decisions will be. The salient point is this: *Knowledge and ignorance compose the independent principle of war, upon which all other principles rely for application.*

Whatever balance we attain between knowledge and ignorance will affect the application of all other principles. In a given scenario, if we have great knowledge, we will tend more toward command than toward anarchy, more toward distribution than concentration, and more toward activity than toward security. If in another scenario, we are largely ignorant concerning the battlefield, we will tend to favor reaction over opportunity, confrontation over dislocation, and objective over option acceleration. All of the dependent principles of war are *knowledge-based* or *ignorance-based* in application.

Because of this special relationship between the principle of knowledge and ignorance and the other principles, we consider this principle to be independent and the others to be dependent.

The principle of knowledge and ignorance gives rise to three categories of dependent principles: principles of aggression, principles of interaction, and principles of control.

PRINCIPLES OF AGRESSION

The principles of aggression deal with what we intend to do to the enemy to accomplish our goals. These two principles are concerned wholly with the means of defeating the enemy.

The Principle of *Dislocation and Confrontation*

Dislocation is the art of rendering enemy strength irrelevant. Confrontation is the direct, symmetrical engagement of enemy strength. Effective warfighting rests upon the skillful combination of dislocation and confrontation.

Dislocation is asymmetrical, and, according to the dual nature of conflict, it is objective in its approach to warfare. Dislocation is an extremely economical means of defeating the enemy, because it sets aside the enemy's strength, rather than expending time, lives, and treasure to destroy it.

Dislocation comes in four forms: *Positional* dislocation seeks to render the enemy's strength irrelevant by causing it to be out of position, oriented the wrong way, or in bad terrain. *Functional* dislocation renders enemy strength irrelevant by causing a key element of it to be dysfunctional. *Moral* dislocation causes enemy strength to be irrelevant, due to the unwillingness of his soldiers or leaders to fight. *Temporal* dislocation renders enemy strength irrelevant through the manipulation of time, attacking the enemy when he is unready.

Confrontation is the subjective part of warfare. It is extremely uneconomical. Confrontation seeks to account for enemy strengths and intentions. Because warfare is competitive, it is illogical to assume that the friendly force will permit continual and uninterrupted dislocation of his strength. Rather, he will oppose us vigorously and continuously. Therefore, we must confront his strength in order to facilitate our dislocation activities.

The purpose of confrontation is to immobilize, delay, and attrit the enemy's strength. We achieve this effect primarily through symmetrical engagement.

This principle is dependent upon knowledge and ignorance for application. The greater our knowledge, the more likely we can dislocate enemy strength. The greater our ignorance, the more likely we will be forced to confront enemy strength. Therefore, every gain of information should lead to greater emphasis upon dislocation.

The Principle of *Distribution and Concentration*
Concentration involves the garnering of combat power with a view to application of that combat power in a specific place and time. Because concentration occurs in both space and time, there are both spatial and temporal opposites to concentration. Spatial distribution means the physical dispersion of forces throughout a given area. Temporal distribution means the separation of forces in time, according to velocity and acceleration.

Spatial distribution is the active dispersion of combat power according to purpose throughout the battlefield, theater of operations, or theater of war. Distribution apportions combat power to accomplish specific purposes in the most economical and precise way possible. Where there is no purpose, there are likewise no forces. Where there is a purpose to accomplish, there is just enough force to accomplish it, without wasteful excess against uncertainty. (This is, admittedly, a revolutionary concept—unthinkable in the context of mass and ignorance, upon which all of our current notions of warfare are based.)

Temporal distribution (also known as preemption) is the *temporal* converse of concentration. Preemption sacrifices combat power to achieve a temporal advantage, with a view to attacking an unready enemy. Concentration sacrifices time in order to garner combat power, with a view to attacking a ready enemy.

Preemption requires speed, in order to hasten contact with the enemy before he is ready. This relationship between preemption and speed is instructive. Obviously, if we desire to conduct preemptive warfare, we know that we will have to increase our speed. But equally as important, this principle instructs us that *increases in speed should lead to preemption.* Using speed merely as a way to hasten attrition fighting is wasteful.

The principle of distribution and concentration is highly dependent upon knowledge and ignorance for application. The greater our knowledge, the more effectively we can distribute combat power. The greater our ignorance, the more we have to concentrate in order to compensate for uncertainty. This applies to every level of war. The inherent ignorance of dumb munitions requires compensation through a high volume of attacks. Conversely, very smart munitions allow for fewer numbers to achieve desired effects. At the strategic level of war, ignorance requires a concentration of greater numbers of troops as a hedge against the unknown. Knowledge, on the other hand, permits economical distribution of precise combat power to specific purposes.

A proper and effective balance between concentration and its two converses, spatial and temporal distribution, leads to success. An overreliance upon either extreme will just as certainly lead to disaster, because the enemy will react to diminish the effects of either. The commander in war should structure his plan to alternate between distribution and concentration. This balance is an important correction to some classical writers' insistence on always remaining concentrated for battle.

PRINCIPLES OF INTERACTION

The principles of interaction address the interplay between the friendly and enemy force. These two principles acknowledge that the enemy is determined and capable, and that our warfighting must account for his aggressive actions.

The Principle of *Activity and Security*

Security consists of those measures taken to protect the friendly force from enemy action. Activity (in this context) is defined as all other friendly action that advances the commander's plan (i.e., other than security). These two are opposites, because the commander has limited resources of time, soldiers, and supplies. He can allocate resources to either security or activity.

The law of economy bears directly on this principle. The goal is to allocate precisely enough resources to security to counter enemy attacks . . . and no more. The commander should seek to reserve as many resources as possible for activity, because it is through activity (moving, fighting, controlling, etc.) that he will prevail in conflict.

Knowledge and ignorance condition the application of the principle of activity and security. The greater our knowledge, the more economically we can secure ourselves. The greater our ignorance, the more we must secure against the unknown.

The Principle of *Opportunity and Reaction*

Opportunity is the freedom to act. The commander seeking opportunity increases his logistical strength, his force's physical mobility, and his army's political penetration into the theater of operations. The garnering of opportunity results in a multiplication of options for positive activity. In short, an army with opportunity has great freedom of action.

Reaction aims at the destruction of enemy opportunity. It takes cognizance of the fact that at times the enemy will have the opportunity to act against us. Reactive warfare therefore accounts for enemy freedom and mobility and attempts to control it, limit it, and eventually destroy it, thereby recapturing opportunity for the friendly force. In the past, armies have often relied upon the mystical notions of offensive operations and initiative as the best way to recapture opportunity. But classic concepts of initiative exist and have meaning only in the context of ignorance. The more pervasive knowledge is, the less apparent and relevant this type of initiative becomes.

The principle of knowledge and ignorance bears directly upon this principle. When an army in a conflict has great knowledge, opportunity is the dominant form of warfare. Knowledge-based armies should spend most of their time exploiting opportunity. When an army has great ignorance, reactive warfare is the norm. Ignorance-based armies will spend most of their time reacting and trying to create opportunity, sometimes through the use of risky offensive actions (in accordance with the old principle of offensive). Modern armies must develop and nurture a strong balance between opportunity and reaction. They must be adept at exploiting opportunity when they have it, rather than frittering it away in idleness and inertia. Conversely, they must be skilled in *creating* opportunity through the prosecution of reactive warfare, that is, through the destruction of enemy opportunity.

PRINCIPLES OF CONTROL

The principles of control address how we manage the friendly force. These two principles acknowledge that the methods we use to control our forces impacts on their chances of success in conflict with the enemy.

The Principles of *Option Acceleration and Objective*

Option acceleration seeks to delay the decision concerning the desired end state of a conflict, and then capitalizes on flexibility to achieve a precise and high-payoff end state. The commander fighting according to this concept uses combat power to rapidly create tactical, operational, and strategic options at a rate that overturns enemy plans and reactions. The strengths of option acceleration are agility in action, and the ability to exploit unforeseen opportunities.

Objective seeks to make an early decision concerning the desired end state of a conflict, and then capitalizes on that decision through a rapid and focused campaign. This concept unifies action through the selection of a stable and achievable goal prior to combat operations. The strengths of objective are the ability to prepare thoroughly, to stay unified and focused throughout the force, and to garner political will.

The principle of knowledge and ignorance bears on the application of this principle. The greater knowledge the superior authority has (whether governmental at the strategic level of war, or military at the tactical/operational level of war), the greater the potential for option acceleration. Conversely, the more ignorant the superior authority is, the more it must rely upon objective.

The Principles of *Command and Anarchy*

Command is the legal and procedural exercise of authority over subordinates. Anarchy calls for the orchestration of the activities of separate entities. On the extreme command side of this principle, we have a well-defined hierarchy, institutionalized with accepted rules and practices. On the extreme anarchy side, we have coequal, independent bodies with no legal or procedural connections at all. Obviously, real future scenarios will feature some combination of these conditions.

Command seeks unity of effort through authoritative direction. Although it tends to gain from efficiency of control, it tends to lose efficiency of interaction with the enemy. The command side of this principle leads to rapid, economical decision making, but it suffers from imposing uneconomical constraint upon the activities of subordinates.

Anarchy seeks success through skillful integration of effects. Anarchy tends to gain efficiency of interaction with the enemy, but it loses efficiency in control processes. Anarchy leads to economical optimization of subordinate activities, but it suffers from uneconomical decision making.

The principle of knowledge and ignorance conditions the application of command and anarchy. The greater the knowledge of the higher headquarters, the more it can and should effectively employ command. The greater the ignorance of the higher headquarters, the more it can effectively use anarchy. A disruption of control occurs when an army fails to balance the principle of knowledge and ignorance with the principle of command and anarchy. For exam-

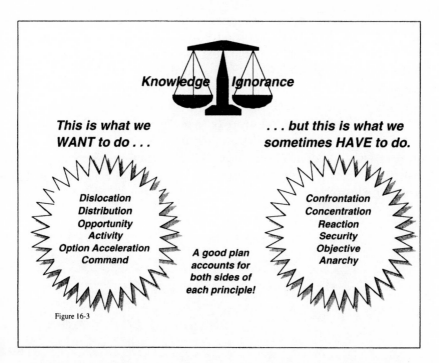

Figure 16-3

ple, when an ignorant higher headquarters tries to command (i.e., employ rigid centralized decision making), failure results. Conversely, when a higher headquarters with great knowledge chooses to orchestrate anarchy, opportunities are missed. Command should accompany knowledge at whatever level it is found. Likewise, anarchy should follow ignorance.

Conclusion

The principles of war, when properly used as arguments rather than aphorisms, provide a dynamic framework for the development of creative solutions in conflict. They adapt to the uniqueness of new situations, while at the same time benefiting from past experience and theory.

Determined application of the principles of war, underpinned by the laws of war, will result in success. We know from conflict theory, however, that the enemy will adapt to any apparent success. The greater our success, the more rapidly he will adapt. Therefore, the principles are designed to account for the phenomena of adaptation and diminishing effects. The balance found among the seven principles must be renewed and revalidated every day, in every battle.

Future warriors must synthesize creative solutions in tomorrow's operations. Rote memorization and slavish obedience to the classical principles of war will no more lead to success than will carrying a rabbit's foot into battle. Victory against a determined enemy requires clear and comprehensive thinking and acting. These revised principles of war provide the framework; character and courage must create the victory.

Appendix: The Epistemology of the Principles of War

As we reevaluate and revise the principles of war, we must frame the discussion. We could begin with an exhaustive study of the semantic origins of the principles. We could reach back to Sun-tzu and the ancient Greek historians or even further to elicit the first expressions of military prescription. From there, we could trace military thought through the golden age of Rome, comparing occidental philosophy with the oriental. The earliest roots of European military traditions, springing from the Dark Ages into the Middle Ages, and thence into the Renaissance could consume several chapters. Parallel to this effort, we would have to compare feudal European concepts of war with the long development of Chinese thought from the Shang, Chou, Han, Tang, and Sung Dynasties. Meanwhile, we would have to account for early martial principles from other corners of the globe, most notably India, Africa, and the Americas.

Such a study would then blossom much more fruitfully (in terms of extant sources) as we emerge into the Renaissance, the Enlightenment, and the early modern period, where the principles as we have them today began to solidify—or even ossify, depending upon one's point of view. In short order, we would find ourselves immersed in the world wars and in the intellectual miasma of Atomic Age military theory.

Indeed, it would be impossible to progress in a serious study of the principles without such a background. It is important to learn what each notable from the past thought about this or that aspect of war.

Nevertheless, this book does not offer such a complete audit trail of thought through the ages for three reasons: First and foremost, previous writers have already dealt with the subject sufficiently, most notably John I. Alger in his book *The Quest for Victory: The History of the Principles of War*. Second, a repetition of these facts would steal space away from a much more urgent effort—to wit, a complete evaluation and revision of the principles. Third, historical works on the principles, necessary as they are, tend to gravitate around semantics, rather than around epistemology.

AN EPISTEMOLOGICAL APPROACH TO
THE PRINCIPLES OF WAR

I will admit that, as a professional infantry officer, when I first heard the word *epistemology*, I thought it had something to do with field sanitation. I certainly never imagined that, later in my career, it would become a subject of urgency to me. But it seems to me that if a soldier ever reaches the point at which he thinks seriously about the military profession, he must begin to consider—and later to challenge—the intellectual origins of the theories that underlie the practice of arms.

In our daily lives, we hardly ever contemplate the rather intriguing force of gravity: how it works, where it came from. Throughout our lives, and those of our ancestors, it simply has always been there. It is an unchanging, physical factor in the existence of mankind. But the theories and doctrines of arms that likewise condition our existence today are not so stable. The factors which govern organized violence change at every sunrise, and although many great thinkers of the past have advanced the art and science of arms, we cannot, upon reflection, relax in their conclusions. Instead, we must—if we think seriously about the world around us—admit to a little nervousness when we receive instruction from the ancients, who lived in a world very different from the one we live in.

But through epistemology—the study of the nature and origins of knowledge—we may begin to get our arms around military truth as it applies to our world today and tomorrow. As we apply ourselves to understanding what has been said about war (and why it was said), we will come away with two products. First, as mentioned previously, we will have a clearer understanding of the semantic history of the principles of war. Second, we will have a more general and categorical understanding of the conceptual issues that framed the arguments of past thinkers.

The purpose of this appendix is to accomplish the latter. In a few short pages, we shall attempt to provide a comprehensive though concise explanation of the four major arguments that have led to the principles of war as we have them today in the United States. This explanation shall be *categorical* rather than *chronological*. If I were to teach a complete course on the principles of war, I should insist that the students first ground themselves in the chronological history of them, because an evaluation of the principles that is divorced from history is worthless. But once the student had progressed through this preliminary study, it would be time to advance into categorization and evaluation. Otherwise, the academic preparation is superfluous.

As the reader studies this appendix, then, the intent is to go beyond didactic, strictly factual assessment. In Part 2 of this book, we took on each of the principles. But the bridge between historical data collection and the somewhat iconoclastic revision of Part 2 is a general conceptualization of what military thinkers have written and thought about the principles of war.

THE FRAMEWORK

There are four major issues surrounding the development and use of the principles of war: their *existence*, their *reduction*, their *scope*, and their *applicability*. It is pointless to delve into a detailed study of the principles without understanding these four great issues. Conversely, by comprehending them, the student of war can easily understand, relate, and compare what the ancients have said about the military art.

The Existence of the Principles of War

Modern inquiries into the principles of war almost always take on the most fundamental question of all: Do they exist? In virtually all literature related to an evaluation or even a recitation of the principles, that question comes up.

There are distinguishable factors in epistemology that would lead to the conclusion that there are no viable principles that govern warfare. From the long tradition of warrior kings comes a semireligious notion that warfare was the purview of the divinely chosen ruler. Just as a commoner could not think of questioning the right of a king to rule, likewise the humble subject could not deign to inquire into the art and science of war, because warfare was a holy and secretive act—

a matter for kings, not for peasant farmers. Even if there were "principles" governing such an enterprise, there would be no point in committing them to writing or to teaching them. Knowledge of such matters connoted authority. Just as ownership and possession of arms has, from time to time, been restricted to the ruling classes, so the right to think about warfare was reserved for the privileged few.

This extreme, mystical view of warfare could not last the scrutiny of the ages. If commoners feared to question their kings about their decisions in war, historians did not. Thus we have Herodotus, Thucydides, Xenophon, Polybius, and others discussing and evaluating the military decisions of the rulers of their day. Long before the divine right of kings was seriously challenged, men began to inquire into the practice of arms. Kings could make mistakes, and the art of war could be learned.

But even in modern discourses on war, one can find prominent writers and thinkers who disallow the idea that principles of war exist. Mao Tse-tung, founder of the People's Republic of China, insisted that each war is different, and that it is a capital mistake to import lessons from a past war into a future one, because the factors conditioning each conflict are utterly different. Likewise, Helmuth von Moltke the Elder, chief of the Prussian general staff from 1858 through 1888, resisted the school of thought that espoused military principles. Having led the German state through the Wars of German Unification, Moltke had applied himself and his staff to the detailed military problems of his day, and he eschewed the proposition that there were general rules that could govern military practice.

In our own day, Michael Howard, among others, has argued that the so-called principles of war exist only as "a crutch for weak minds." Bernard Brodie, a prominent writer on strategic matters in the nuclear age, is likewise held to disbelieve in their existence. (We shall deal with Brodie's actual thinking on the subject following.)

The bottom line is that the student of military doctrine will find a school of thinking—generally small but vocal—that argues there can be no principles governing warfare, because each situation is unique. Hence, in the purest sense of this viewpoint, we can learn no applicable lessons, nor derive any stable truths from past military events.

But the claim that there are no principles to be learned is paradoxical. One must question the deceased, such as Moltke and Mao, on the logic of writing their respective books if principles do not exist. Does not the very belief in the nonexistence of principles count as a principle itself? And what is the purpose of writing one's thoughts on any subject if we truly believe that knowledge cannot be usefully categorized, learned, and taught? It's like writing a book about the nonexistence of books, or speaking about the uselessness of speech.

We should not be surprised, then, to discover that most writers and thinkers of the past believed that there was a body of truth about war that can be learned and put to some use. As parents, we can appreciate the uniqueness of each child and yet apply the experience of raising one to the raising of another.

In order for us to carry on with our study, we must take the viewpoint that it is possible to develop *principles*—however we define that word—to assist us in the study and practice of conflict in general and armed conflict in particular.

The Reduction of the Principles of War

> There are but few general principles for the conduct of armies, but their application calls forth a host of combinations which can neither be foreseen nor classified as rules.
> —Marmont

If we proceed past the first argument and agree that there *are* principles, we must next advance to the second question: Can the principles be reduced, enumerated, and expressed? In other words, given that there is a body of truth concerning warfare, is it possible and advisable to reduce that truth into writing? Can a student of war recognize discrete principles and express them as a finite list of words or aphorisms?

On the surface, it would seem a small matter to go from a belief in the *existence* of principles to a simple listing of those principles. And yet, the present form of the principles of war is a fairly new phenomenon, as the semantic histories would attest. In this, the principles of war have been analogous to the concept of morality. Although most people would agree that morality is important, the arguments would begin in earnest when they tried to reduce the concept of morality into a list of moral rules.

Likewise, most writers of the past agree that one can learn about war. Most even agree that there are principles that are fundamental to understanding and practicing warfare. But when it comes to actually enumerating those principles, the literature on the subject does not offer much.

Therefore, on the question of the reduction of principles, we will find both arguments in the negative and the affirmative. On the negative side, we will see that there are some writers who refuse by omission to list the principles, although they profess belief in them. We will also find a handful of modern writers who assert that principles of warfare are too complex to be reduced to one-word truths or proverbs.

Today, young officer cadets can make use of handy little acronyms as memory aids to help them recall the nine accepted principles of war. Typically, students use the cacophonous "MOSS-MOUSE" or some other variation as a mnemonic device to remember "Mass, Objective, Surprise, Simplicity, Maneuver, Offensive, Unity of Command, Security, Economy of Force." Whereas great thinkers of the past, like Ferdinand Foch, sometimes wrote whole treatises on the principles of war and yet never reached an actual listing of those principles, today's American soldiers, airmen, sailors, and marines can reduce the mystical principles to a single acronym.

But to summarize the reduction argument, the reader will find a clear trend. Most past writers, from the ancients through the twentieth century, simply didn't try to reduce the principles into a list of any sort. Typically, military writers—whether historians, philosophers, or soldiers—believed in the existence of truth about war, but they avoided (and perhaps saw no need for) an enumeration of discrete, finite aphorisms. There is no doubt, however, that certain general truths emerged from generation to generation. Thus, one may not find a principle called "Morale" listed in Xenophon's *Anabasis*, or Sun-tzu's *Art of War*, but both authors offer incisive and colorful instruction on the subject.

It was only in relatively recent times that both students and teachers of military history and theory began to require a reduction of the principles to a written, comprehensible list. The renewed emphasis upon science, scientific method, and technical precision that came about in Western culture through the nineteenth and twentieth centuries no doubt led to a new scrutiny of the "science" of war, and no science can exist without principles.

From the time of Jomini and Clausewitz, we can find a growing trend to reduce the principles governing warfare to a teachable list. As the semantic histories will show, the mystical truths about war that were handed down from the ancient world through the Middle Ages eventually took shape into sentences and discrete aphorisms. Finally, J. F. C. Fuller, a prominent and controversial British soldier and thinker, reduced the principles into a list of eight and later nine principles: Direction (or Objective), Offensive Action, Surprise, Concentration, Distribution, Security, Mobility, Endurance, and Determination. Fuller also considered Economy of Force to be a "Law of War."

From this initial list, the principles found their way eventually into American military doctrine. There were some changes from Fuller's prescriptions—some substantive, most not—and in 1921, the U.S. Army first published the principles of war in its written doctrine.

Almost as soon as the principles of war were reduced into a listing of sentences or words, military professionals balked, both in our country and abroad. This kind of argument is healthy and necessary, and the course of it generally revolved around which truths were important and immutable enough to be principles, and which were not. But from the start there have been skeptics who resisted any attempt to reduce military truth into military *truths*.

In the Atomic Age, Bernard Brodie, a well-respected strategist from the Rand Corporation, argued against the principles of war being taught in the U.S. armed services. In a lecture at the Command and General Staff College in 1956, Brodie made it clear that he had no problems with the notion that military principles existed, but that he did not accept the reduction of those principles into aphorisms or single words. He believed that past writers on the subject considered the "principles" to be "well understood, commonly accepted philosophy . . . on the governance of strategy." The problem was that now those principles had become a specific body of axioms.

Strategy, according to Brodie, is the oldest but least developed science, born long before the development of modern science. Traditionally, military strategy was nonintellectual, nonanalytical, and unscientific. Through the years, as previously outlined, successful strategic formulation evolved into axioms, and later into very abbreviated aphorisms. Brodie characterized the modern age as one which uses condensation as a way of life, but the principles of war are too condensed.

What has been lost in the condensation or reduction of the principles of war? According to Brodie, they became too abstract and too devoid of content. Further, they are subject to exception and qualification. Once codified, they can become a roadblock to thinking, because they are too respected. In the end, Brodie suggested to the students that the principles of war be considered like the titles of chapters in a book. By themselves, these "chapter titles" are of little use; the student gains useful knowledge only by actually reading the chapters. He can then use the chapter titles as a reminder of what the chapter stated.

This second major argument concerning the principles of war will likely continue, and should continue. Indeed, in some respects, our later revision of the principles will revisit the question of reduction. But to proceed in our understanding, we must assume for the moment that (1) the principles exist; and (2) they have been reduced—for better or worse—to a set of nine aphorisms.

The Scope of the Principles of War
The third argument concerning the principles of war is their scope. What do the principles apply to? Although they are called principles of *war*, is that truly their correct scope? Do they provide counsel on war? On campaigning? On battlefield tactics? Do they provide insights on strategy, operational art, or only the tactical level of war? Do they apply equally to nuclear/biological/chemical warfare as well as to conventional conflict? Do they govern counterguerrilla tactics? Peacekeeping? Disaster relief? Do they apply equally on land, on the sea, and in the air?

Modern American military doctrine distinguishes among at least three levels of military operations: strategic, operational, and tactical. The strategic level concerns the use of military resources to attain national political objectives. High-level strategy is all about translating political will into an integrated plan of action. The term "grand strategy" refers to the use of *all* sources of national power, including the military, diplomatic, informational, economic, and intelligence assets of the nation.

The operational level of war, which the U.S. armed forces began to study officially in 1982, deals with the conduct of campaigns and major operations. As the nature of modern war changed in the late nineteenth and early twentieth centuries, American doctrine graduated from a fixation upon individual battles, into an appreciation of the need to convert transitory tactical successes into a meaningful campaign to serve the national strategy. Thus the theme of American warfighting doctrine since the early eighties has been operational-level warfighting.

The tactical level of war concerns the fighting of battles. Tactical instruction in the U.S. Army aims at producing soldiers and leaders who can best employ their weapons to defeat, destroy, or capture enemy forces through fire and maneuver.

This threefold approach to warfighting creates a problem for the student of the principles of war. One of the basic assumptions underlying the distinctions among the levels of war is that an action or event that might be good from one perspective might be bad from another. For example, if a corps commander commits a light infantry brigade to an air assault to help seal off an enemy in a particular battle, we would, from a tactical perspective, applaud the battlefield advantage thus gained. By adding to our forces in battle, and by seizing a critical piece of terrain, we have gained the tactical advantage over the enemy. However, the light brigade used to air assault into the battle is now not available to seize the lightly defended bridges over the next major river that the corps must cross. Thus, from an operational perspective, the commitment of the light brigade to battle was a setback.

It is by understanding the potential tradeoffs among tactical, operational, and strategic considerations that the student of war learns the complex nature of these various levels. But if there are distinctions and even tension among the levels of war, then how do the principles of war fit in? What level of war do they truly pertain to?

The simplest way of dealing with this question is to assume that they apply, in some way, to all three levels of war. But in this case, the simplest solution is also wrong. Instead, we must understand that the principles were derived from experience and writings concerning warfare in the Agrarian and early Industrial Ages. At that stage in the development of military thought, *there was little or no distinction among the levels of war*. From the earliest ancient history through the mid-1800s, military thought developed around the *battle*. Although much had been written on matters other than the battlefield, it is no overstatement to assert that warfare was, in the minds of most practitioners, mostly about fighting and winning a decisive battle.

Hence, in a very real sense, strategy and tactics were, if not wholly synonymous, at least closely related. To quote Marshal Marmont's definition of strategy: "General movements, executed beyond the sight of the enemy and before the battle, are called *strategy*." In this context, we will not be surprised to find that the so-called principles of war appear under close scrutiny to sound a lot more like principles of battle.

Likewise, much of our military past occurred in the context of what at least one author referred to as the "Heroic Age." That is, many past wars have been fought with the understanding that the subject populations on both sides would acquiesce in the battlefield decision between armies. In other words, if one nation's army defeats the other, all sides will, for the most part, agree to the victor's demands. In practice, this arrangement has been much more protracted and complex, but the underlying assumption that the civilian populace will follow the lead of the military combatants in resolving the political dispute pertained.

There have been times in history, however, when civilian populations did not acquiesce. In this event, victorious armies had the double duty of first defeating an opposing armed force, and then supressing the subject population. In an age of modern commnications and heightened political sensitivities, this trend of divorcing political resolution from battlefield victory must increase—and does. The result is that the principles, which may have governed almost the totality of war in the past, now must contend with an utterly different problem: how to make tactical success relevant to political outcome. We should not be surprised that principles conceived in the Heroic context offer little or nothing on this problem.

Finally, the principles of war were discovered, reduced, and enumerated long before the invention of flight. Since so much of modern naval operations, as well as air operations, depends upon the use of air power, we are justified in asking ourselves whether the principles of war can govern conflict at sea and in the air. Again, we will find a considerable body of literature that takes the easy approach: The principles apply to all forms of armed conflict. But the relationships among conflict on the land, on the sea, and in the air bear close scrutiny. Beyond the most general and conceptual level of analysis, we find that operations in these various media are indeed dramatically different from one another.

To proceed from this argument, we can temporarily propose that the scope of the principles of war is (1) focused originally at the tactical level of war, although it may have some application at other levels; (2) concerned mostly with conventional armed conflict between relatively symmetrical opponents, although there may be some applicability in other scenarios; and (3) confined mostly to land warfare, although the logic of the principles may extend with modification to other media. This listing of assumptions about the scope of the principles of war is intended only as a vehicle to continue our analysis, and the reader is free to disagree.

The Applicability of the Principles of War

A genius applies recognized principles; in this consists the whole art of war. —Marmont

Finally, we must examine the fourth major argument concerning the epistemology of the principles. How are they to be applied or used?

To begin with, we must admit (if we are honest with ourselves) that the principles are used to evaluate the past far more often than they are used to predict and ordain the future. It is a simple matter to regard a past victory and twist the historical record into an affirmation of the vaunted principles of war. It is altogether another matter to use those principles to confidently create and foresee victory in the future. Nevertheless, if the student of war never graduates to this latter application, then his study of the art must be considered ultimately just an academic exercise.

The correct use of the principles of war—in whatever state we find them at the end of this book—is to create a better future. Scholarly meditation, an enjoyable recreation though it may be, offers no real test of the utility of the principles.

But beyond this issue, the applicability question concerns what the student or practitioner of warfare is to do with the principles of war. This question is answered in roughly three different ways by past and contemporary writers. First, there is the extreme view that the principles of war (or their unreduced antecedents) must be strictly observed and obeyed at all times, lest disaster and defeat result. We may smile at Sun-tzu's warning that "those who master them win; those who do not are defeated." Or, as he advised his sovereign: "If a general who heeds my strategy is employed he is certain to win. Retain him! When one who refuses to listen to my strategy is employed, he is certain to be defeated. Dismiss him!"

Likewise, one may find a small but amusing group of writers who believe so passionately in their conclusions that they can permit no deviation or further scrutiny of their rules. But didactic teaching and dogmatic insistence on utter obedience does not bear much fruit in today's society. Hence, we usually find that most writing on principles of warfare recognizes the need for flexible application.

The requirement for flexibility derives from two factors. First, as noted in the first argument, every military situation is different. Therefore, the theory and practice of arms must adapt to each unique occasion. But the other factor requiring flexible application is that the principles of war themselves in some ways contradict each other.

Mass and maneuver, for example, can trade off against each other. A massed army is difficult to maneuver, because its fighting formation and its overcrowding of local road networks slow its movement. Conversely, an army properly disposed for rapid maneuver is typically dispersed along movement routes and in formations adapted for speed. This simple theoretical trade-off has translated many times in history into sudden, bloody ambushes that left many men dead, while the remaining few were left to ponder the tension between movement and massing for battle.

Mass also trades off against security. In apportioning combat power, the commander must always find the balance between securing his own vulnerable rear and flanks with the need to mass combat power against the enemy to his front.

Mass can even trade off against unity of command. During the Gulf War, the United States opted for the building of a multinational coalition in order to create mass—both in the military and political sense. This amalgamation of various national armies resulted in a command structure that was at best difficult and at worst highly volatile. In the end, history proved the choice to be a wise one. Others in history have not been so fortunate.

During the Boxer Rebellion at the turn of the century, the U.S. Army found itself in common cause with British, French, Japanese, German, and Russian allies (along with a few other minor participants) against a peculiar array of Chinese army contingents and the mystic militia of the Boxers in the Shantung Province of China. Technological advantage and the most incompetent Chinese government imaginable permitted the allies a victory of sorts. But throughout the campaign to relieve the besieged legations in Peking, the allies blundered time and again through misunderstood orders, rampant mistrust, and deliberate treachery. There is little doubt that if faced with serious opposition the shaky coalition would have fallen apart.

In a similar manner, each of the principles tend to trade off against the others, thus requiring great flexibility of thought and application.

The final phase of the applicability argument proposes that the principles of war should not be viewed as *aphorisms* or *prescriptions*, but rather as *arguments*. As I concluded in Chapter 15, this is, in my view, a much more effective way to deal with the principles. Employing dialectic logic, we eschew any attempt to formulate aphorisms for a future conflict that we can scarcely envision, let alone control. Instead, we simply define the categories of thinking that a future soldier must consider, and we define the end points of the possibilities available to him. We then let him free to synthesize the right answer to meet his particular challenges.

Conclusion

With a prudent respect for the past, we must be intellectually ready to question every assumption and assertion, because people will die and societies will suffer violence and destruction at the hands of these principles. To the degree that we can scrutinize the wisdom of the past, we may ordain a better future.

Works Cited

Alger, John I., *The Quest for Victory: The History of the Principles of War.* Westport, Connecticut: Greenwood Press, 1982.

The definitive work on the history and background of the principles of war. Alger painstakingly outlined both Eastern and Western thought on the principles from ancient times down to today. An invaluable addition to the literature on the subject.

Campen, Alan D., *The First Information War.* Fairfax, Virginia: AFCEA International Press, 1992.

This book offers a peculiar perspective on the Gulf War of 1990–91 as being the first war in history to rely so heavily on information technology. To the degree that the reader can stomach technical jargon and detail, this is a useful look at some of the complexities of modern warfighting. After two complete readings, I can understand about two-thirds of the book.

Clausewitz, Carl, *On War.* Princeton, New Jersey: Princeton University Press, 1976.

One of the great cornerstone works of Western military thought. Over the years, I learned far more from reading Clausewitz directly than I ever learned from the writings and lectures of so-called experts on the master. Much of the instruction on Clausewitz that I received as an army officer was pedantic, argumentative, and ineffective. Many commentators on Clausewitz have missed the tremendous application of his thought to future warfare, fixed as they are on endless semantic debate. Read Clausewitz; he is smarter than his interpreters.

Corbett, Julian S., *Some Principles of Maritime Strategy*. London: Long-mans, Green, and Co., 1911.

This has been one of my favorite books since I first picked it up. Strangely, a book about turn-of-the-century naval warfare was where my first lesson on the duality of conflict came from. Corbett's explanation of the complementary roles of ships of the line and cruisers is of direct and immeasurable value to today. Although the technology has changed, the logic remains.

Dupuy, Col. T. N., *Numbers, Predictions, and War: Using History to Evaluate Combat Factors and Predict the Outcome of Battles*. Indianapolis: The Bobbs-Merrill Company, Inc., 1979.

Dupuy's contribution to military history is almost unparalleled in the United States, and his historical works are among my most oft-used resources. His theoretical work, on the other hand, is useless, except to show how wide a gap there is between history and the correct use of history. Dupuy's two books on the subject of his Quantified Judgment Model are strewn with errors of logic that the author tries to recoup with intellectual arrogance and assertion. Following his gross overestimation of expected casualties in the Gulf War, his model and method have not been heard from. Nonetheless, this book is superb instruction in how not to interpret historical data.

Foch, Gen. Ferdinand, *The Principles of War*. New York: AMS Press, 1918.

Although the student of war must admire Foch's circumspection and intellectual bent, his book on the principles of war borders on the incoherent. He seems most impressed with Clausewitz and Goltz, and with Napoleon, who, in Foch's estimate, provides the most enduring example of how to make war. Of the principles, the reader will learn very little, except concerning the author's repeated assertion that they exist.

Freedman, Lawrence, and Efraim Karsh, *The Gulf Conflict 1990–1991: Diplomacy and War in the New World Order*. Princeton, New Jersey: Princeton University Press, 1993.

Fuller, Col. John Frederick Charles, *The Foundations of the Science of War.* London: Hutchinson & Co., Ltd., 1926.

One of the most profound books on military science, Fuller's work is a difficult read. His concepts are at times a reflection more of his own mystical beliefs than of science, but his brilliant intellect, passionate arrogance, and broad educational background combine to make this book a treasure to read. A unique and useful work.

Johnsen, William T., et al., *The Principles of War in the 21st Century: Strategic Considerations.* Carlisle, Pennsylvania: U.S. Army War College, 1995.

This small book is an excellent attempt at revising the principles of war. The authors' thinking is well categorized and precisely scoped to the strategic level of war. Although the intent is to be open-minded in their approach, the writers begin with the assumption that the principles are existent and valid. Beginning with the basic nine, they find it difficult to stray too far from home. However, the book certainly is a good start at revision.

Kingston-McCloughry, E. J., *War in Three Dimensions: The Impact of Airpower Upon the Classical Principles of War.* London: Jonathan Cape, 1949.

A very worthwhile though short piece. The author demonstrates the ability to evaluate airpower in the greater context of armed conflict and even grand strategy. The book contains some thoughtful analysis, despite the dated technological references.

Kotenev, Capt. Anatol M., *The Chinese Soldier: Basic Principles, Spirit, Science of War, and Heroes of the Chinese Armies.* Shanghai: Kelly and Walsh, Ltd., 1937.

Written prior to the influence of Mao Tse-tung, this pre–World War II work confirms the salient influence of Sun-tzu on Chinese military thought. Although it details many other influential thinkers, soldiers, and writers, each is viewed in relation to the ancient master.

Lendy, Capt. A. F., *The Principles of War.* London: Mitchell and Son, 1862.

Despite the title, this work has nothing to do with the principles of war, addressing, instead, the doctrinal framework of warfare in the mid-nineteenth century. The author provides an edited rendition of terms and definitions concerning mostly the linear battle tactics of his day. His paradigm is decidedly Napoleonic, with consequent fixation upon the battle.

Livy, *The War with Hannibal.*

Luttwak, Edward N., *Strategy: The Logic of War and Peace.* Cambridge, Massachusetts: Belknap Press, 1987.

In this and all his works, Luttwak demonstrates his status as one of the greatest American military thinkers. His power of creative thought, combined with a profound expertise in both contemporary military affairs and history, has defined a solid and resilient body of effective theory. A fine work that deepens the understanding of the dialectic nature of warfare and is absolutely foundational to my understanding of combined-arms theory.

Polybius.
In accordance with what Marshal Marmont once wrote, I found Polybius and Vegetius more curious than useful.

Reid, Brian H. (ed.), *The Science of War: Back to First Principles.* London and New York: Routledge, 1993.

An excellent collection of thought-provoking essays concerning future warfare. Both broad in terms of subjects covered, and deep in its sophistication, this book is well worth reading.

Thucydides, *History of the Peloponnesian War.*

One cannot read Thucydides without being impressed with the subtlety of political thought in ancient Greece. Through his narrative of the war, the author provides incisive instruction in the art of strategy.

Voysey, Lt. Col. R.A.E., T.A. Res., *An Outline of the Principles of War.* Diss, Great Britain: Diss Publishing Co. Ltd., 1934.

An interwar period study of the principles of war as expressed in the British Field Service Regulations. The author contributes no new ideas and instead illustrates the efficacy of existing principles by an examination of the campaigns of the nineteenth century and early twentieth century. As this work looks only backward in time, its tone is didactic and its utility limited.

West, Dr. Joseph (trans.), *Principles of War: A Translation from the Japanese.* Fort Leavenworth, Kansas: U.S. Army Command and General Staff College, 1983.

This work contains some fascinating ideas on military strategy and tactics, which the reader will find refreshingly different from traditional Western thought. Nevertheless, this work suffers from a difficult translation and is hard to read.

Xenophon, *The Anabasis.*

A lively and intriguing story about, perhaps, the most famous retreat of the ancient world. In the few instances in which the author writes of military principles, his thesis is that through imaginative planning and clever maneuver, he and his fellow leaders could avoid hopeless attrition fighting. Although his tactical ideas were not well developed, Xenophon clearly prized forethought over heroism.

Index

Activity and Security, 163, 167, 169, 253, 255, 258

Adaptation, 66, 68–74, 77–79, 136, 211, 219–21, 261–62

Advanced Warfighting Experiments (AWEs), 40, 46, 49, 105, 168, 177–78, 224–25

Aggression, Principles of, 255–57

Air operations, 6, 18, 25, 29. *See also* Joint Surveillance and Target Acquisition Radar System (JSTARS)

Alexander the Great, 23, 158, 160, 233

Anarchy. *See* Command and Anarchy

Antietam, Battle of, 36, 39, 43–48, 200–01

Argument. *See* Dialectic logic

Armor. *See* Infantry, Tanks

Art, 7, 24, 62, 77, 125–26, 130–31, 135, 209, 266; and science, 19, 175–76, 207, 209, 235, 264, 269

Artillery, 4–5, 25–26, 28, 62–64, 67–68, 70–71, 78, 111, 210, 220, 249

Attack profile. *See* Adaptation

Attrition, 5–6

Auftragstaktik. See Command and Control

Aviation, 67, 70, 133

Balance. *See* Dialectic logic

Battle of Britain, 165–66

Chamberlain, Joshua, 85–88

Chess, 91–92, 150–51

Civil affairs, 23, 33, 96, 141

Clausewitz, 32, 84, 94, 140, 158–59, 164, 217, 222, 226, 243, 269

Combat multiplication, 136–37

Combined arms, 57, 65, 67–75, 123, 136–37, 219–21; definition, 67; and Simplicity, 171

Command and Anarchy, 200–204, 255, 260–61

Command and Control, 14, 16, 18, 30, 60, 62, 109–13, 179, 191; centralization of, 179–80, 199–204; warfare, 22–23, 167–69. *See also* Command and Anarchy, Unity of Command

Communications, 14, 21–23, 25–26, 28, 57, 62, 111, 253; and information management, 177–79; and Objective, 152–58, 160; and Security, 167–69

Computers, 16–17, 23, 26–28, 35, 47, 75–79, 111, 173, 204, 253; Age of, 13, 15, 21, 173–74; security of, 167–69

Control, Principles of, 259–61

Convergence, Principles of, 10–12

Defeat, 33, 64, 121, 211, 225; and destruction, 77–78, 211, 225, 249

Defense. *See* Offensive

Destruction. *See* Defeat

Dialectic logic, 243–50, 261; and balance, 65–67

Diminishing effects. *See* Adaptation

Dislocation, 11, 20, 61–67, 70, 74, 79, 106, 122, 180, 211, 249; and Confrontation, 63, 66–67, 79, 127, 193, 253, 255, 255–56; definition, 64; and Duality, 227, 232; and Economy, 127, 136; and Simplicity, 171; and Surprise, 184. *See also* Combined arms

Dispersion, 95–97, 109–13, 116. *See also* Distribution

Distribution, 120; and Concentration, 118, 193, 255, 256–57

Doctrine, 19, 22–23, 33, 60, 117, 119, 122, 135–36, 140, 253, 271; and command philosophy, 179–80; and Duality, 230, 232, 236; and humanity, 209–11, 213–16; and Objective, 160; and Security, 165; and Simplicity, 170; and simulation, 75–79, 117; and Surprise, 188–89; and technology, 6–7, 26–27, 100; and velocity, 192–93

Duality, Law of, 160, 208, 226–39, 249, 251, 255–56

Economics, 9, 15, 54, 69–70, 125, 130–37, 140, 142, 145, 156, 158–59, 187, 224, 227–29, 231, 246–48

Economy, 31, 66, 124–37, 203–04, 232, 235, 255–56, 260; Law of, 128, 137, 197, 208, 217–25, 249, 251, 253, 257–58; and information management, 177–79; and Objective, 144; and Security, 163–67, 169; and truth, 128–30; and Unity of Command, 197, 199, 203–04. *See also* Economy of Force

Economy of Force, 10–12, 124–37, 139, 182

Estimates, 18–19

Experimentation. *See* Advanced Warfighting Experiments

Fixing, 66, 106–07, 127, 133, 136

Force multiplication, 136–37

Franco-Prussian War, 83–84, 143

Fuller, Gen. J.F.C., 124, 126, 128, 138, 182, 217, 269

Gettysburg, Battle of, 85–88

Global positioning system (GPS), 26, 28

Goltz, 80, 97, 103–04, 106, 164, 234

Haig, Gen. Douglas, 3–6, 238, 248–50

Heroic warfare, 141–42, 147, 272–73

Humanity, 127, 196; and information management, 177–78; Law of, 207–16, 217, 251; relationship to warfare, 7, 14, 32, 126. *See also* Moral factors

Hussein, Saddam, 85, 88–89

Infantry, 4, 20, 22, 67, 70, 73,
    111, 203–04, 230–31
Information, 35–49, 104–06,
    128, 130, 219; Age, 12–22, 74,
    85, 90, 94, 114, 129, 152, 156,
    163, 169, 174, 183, 186–87,
    198–99, 251; and authority,
    197–98, 201, 204; manage-
    ment, 174, 176–79; warfare,
    13, 22–33, 35, 67, 128, 164,
    170, 193, 199, 204, 219
Initiative, 80, 82–93, 254, 258
Intelligence, 18–21, 30, 38, 166
Interaction, 103, 156, 227;
    Principles of, 11–12, 91–92,
    258–59

Joint Surveillance and Target
    Acquisition Radar System
    (JSTARS), 18, 39, 48, 71, 88

Knowledge, 19, 105, 169, 198;
    and Ignorance, 34, 40, 44–46,
    49, 88–93, 126, 128–30, 193,
    218–19, 251–55, 256–61. See
    also Truth
Korean War, 147

Lanchester equations, 117, 119,
    123, 211
Laws of War, 207–09, 251. See
    also Duality, Economy,
    Humanity
Leader development, 40, 67,
    76–79, 105, 119, 219–20,
    238–39, 245–50
Lee, Gen. Robert E., 36–40,
    44–45, 47–49, 85, 87, 212

Levels of war. See Operational,
    Strategic, Tactical, and levels
    of war.
Logistics, 21, 27–30, 77–78, 84,
    96, 174, 203–04, 225, 258
Low intensity conflict (LIC). See
    Operations Other Than War

Maneuver, 4, 8–9, 11–12, 21, 33,
    53–79, 109, 112–13, 120–21,
    141
Maneuver warfare, 60, 180, 201
Maryland campaign (1862),
    35–49
Mass, 8–12, 94–123, 127, 130,
    218–19, 232, 237, 249, 253–54,
    257, 274–75; and Objective,
    141–42; and preemption,
    190–91
Massing effects, 8, 94–95,
    117–19, 123
McClellan, Gen. George, 36, 39,
    43–44, 47, 49, 201
Mechanized forces. See Infantry,
    Tanks
Media, 16–17, 22–25, 31, 59
Military profession, 6, 10, 250
Mission creep, 157–58
Moral factors, 14, 32–33, 57–58,
    65, 77–78, 87, 104, 113–15,
    209–16, 249, 268. See also
    Humanity

Naval operations, 29.
Novara, Battle of, 194–95
Nuclear weapons. See Weapons
    of mass destruction

Objective, 9–12, 127, 138–61,
    194, 233, 255. See also Option

Acceleration
Objective conflict. *See* Duality
Offensive, 8–9, 11–12, 30, 80–93, 232, 258
Operation Desert Shield/Storm, 8, 23, 25, 31, 73, 77–78, 88–89, 111
Operation Just Cause, 8
Operational level of war, 9, 24, 29, 81, 105, 115–16, 223, 270–73
Operations Other Than War (OOTW), 158–60
Opportunity, 43, 47, 91–93; and Reaction, 92–93, 127, 186, 253, 255, 258–59
Option Acceleration, 156–58; and Objective, 253, 255, 259–60
Organizations, 57, 121–23, 188

Perpetual unreadiness, 167, 183–84, 188–90, 192–93, 249. *See also* Surprise
Politics, 9–10, 20–21, 23, 31–33, 57–58, 62, 82–83, 97, 122–23, 131, 187, 196, 212, 216, 222, 231, 233, 254, 258; and Objective, 140–41, 143–46, 158–59, 161
Precision, 105, 108–09, 117, 119, 121–23, 219, 225; movement, 27–30; protection, 30–32; strike, 22, 25–27, 59, 71, 73, 105, 113, 210–11, 233–34
Preemption, 11, 83, 189–90, 257. *See also* Surprise, Velocity
Principles of War, Categories of, 10–12; (Classical), 4–5, 7–12,

49, 59, 81, 125, 127, 140–42, 207, 262; epistemology of, 263–76; (New), 9, 49, 127–28, 208, 244–45, 251–62
Psychological operations, 22, 32–33
Purpose audit trail, 131–35
Pursuit operations, 77, 97

Reaction, 92, 127, 253, 255, 258–59; battlefield. *See* Adaptation, Opportunity

Science, 10, 16, 72. *See also* Art and science
Sea lift. *See* Naval operations
Security, 9, 11–12, 23–24, 127, 162–69, 184, 192, 253–54
Sensors, 17–21, 23, 26, 31, 39, 62, 70–73, 164, 167, 184, 253
Sharpsburg. *See* Antietam
Shock, 102–03
Simplicity, 8, 10–12, 170–81
Simulations, 75–79, 114, 117, 122, 177, 209, 215–16, 224–25. *See also* Training
Situational awareness, 41–43, 46, 111–15, 167; and Surprise, 190–93. *See also* Tactical Internet
Somme, First Battle of, 3–6
Spring Offensive (1918), 5, 213
Strategy, 10, 24–25, 29, 105, 140, 145–46, 157, 228
Strategic level of war, 9–10, 81, 83, 138–39, 140–41, 144, 154–55, 202–03, 223, 257, 260, 270–73
Subjective conflict. *See* Duality

Sun-Tzu, 53, 61, 66, 80, 116, 138, 140, 162, 182, 194, 263, 268, 274; and Duality, 226; and military autonomy, 148–52
Super weapons, 73–75
Surprise, 11–12, 42, 57, 65, 141, 171, 182–93, 195. *See also* Preemption, Velocity

Tactical Internet, 47, 111–15, 168, 178
Tactical level of war, 9–10, 81, 138–42, 188–93, 203–04, 228–29, 270–73
Tactics, 4–5, 10, 24, 27, 73, 105, 110, 120–21, 236. *See also* Combined arms
Tanks, 20, 67–68, 70, 72, 101, 111, 137, 183, 191–92, 236–38
Technology, 4, 9–10, 12–22, 36, 40, 43, 48, 56–57, 64, 71–73, 105–06, 111, 113, 120, 122–23, 253; and doctrine, 6–7, 100; and information manage-ment, 177; and Objective, 139–40, 160; and Security, 163, 167–69; and Simplicity, 173; and Surprise, 183, 186–88, 190, 193; and Unity of Command, 196–98
Time, 24, 57, 65, 157, 183–86, 193, 196, 199, 256–57
Total war, 33, 82–83, 142–43
Training, 30, 58–59, 75–78, 188, 191, 199, 215, 219–20, 236, 238–39, 253. *See also* Simulation
Truth, 19, 34, 36, 40, 42, 49, 90, 105, 128–30, 163, 193, 219, 252, 254. *See also* Knowledge

and Ignorance

Unity of Command, 8, 10–12, 194–204
Unmanned Aerial Vehicles (UAVs), 39, 48, 71
Unreadiness. *See* Perpetual unreadiness
Urban warfare, 20–21, 64

Velocity, 41–43, 64, 107–09, 167, 169, 189–93, 256. *See also* Preemption, Surprise
Vietnam War, 139–40, 143, 145–47, 157, 160, 235

Weapons of mass destruction, 12, 16, 74–75, 97, 219, 221–22
Weinberger Doctrine, 145–46
World War I, 3–6, 14, 82, 103, 119, 213–14, 237
World War II, 20, 25, 58–60, 67, 92, 139, 147, 160, 165–66, 182–83, 197, 201, 237